WIZARDS
of OZ

BRETT MASON is Chair of the Council of the National Library of Australia and Adjunct Professor in the School of Justice at the Queensland University of Technology. He was formerly a Senator for Queensland, serving in the ministry, before being appointed Australia's Ambassador to The Hague and Permanent Representative to the Organisation for the Prohibition of Chemical Weapons. He is the author of *Privacy Without Principle* and co-editor of *Future Proofing Australia*.

'What a story of two mighty trees in a forest of champions: the greatest of "Australia's greatest generation". Brett Mason's sparkling prose weaves their stories brilliantly into a compelling account. Unputdownable!'

**General Sir Peter Cosgrove,
former Governor-General and
Chief of the Defence Force**

'Working in Britain, Australian pathologist Howard Florey (Oxford) and physicist Mark Oliphant (Birmingham) made major contribution in WW2. Available by D-Day, penicillin earned Florey, Ernst Chain (his chemist colleague) and Alexander Fleming the 1945 Nobel Prize for Medicine. The Oliphant team's development of, and his advocacy for, the cavity magnetron enabled the airborne radar that allowed pilots to 'see' at night. From 1950, both men played a major part in establishing the Australian National University, Australia's first research university. These are great stories!'

**Professor Peter Doherty,
Nobel Prize winner in Medicine**

'A fascinating, informative read on a wonderful slice of Australian history.'

**Peter FitzSimons,
bestselling author**

'A compelling read: exquisitely detailed, beautifully written, comprehensively researched ... I couldn't put it down until I turned the very last page.'

Hon Julie Bishop

'A truly great book ... Among the very best of scientific histories I have read anywhere in the world.'

Professor Brian Schmidt AC

WIZARDS *of* OZ

How OLIPHANT and FLOREY helped
win the war and shape the modern world

BRETT MASON

NEWSOUTH

A NewSouth book

Published by
NewSouth Publishing
University of New South Wales Press Ltd
University of New South Wales
Sydney NSW 2052
AUSTRALIA
https://unsw.press/

© Brett Mason 2022
First published 2022

10 9 8 7 6 5 4 3 2

This book is copyright. Apart from any fair dealing for the purpose of private study, research, criticism or review, as permitted under the Copyright Act, no part of this book may be reproduced by any process without written permission. Inquiries should be addressed to the publisher.

A catalogue record for this book is available from the National Library of Australia

ISBN   9781742237459 (paperback)
           9781742238548 (ebook)
           9781742239446 (ePDF)

Internal design Susanne Geppert
Cover design Lisa White
Cover images (front) Mark Oliphant, 1941 (Lawrence Berkeley National Laboratory); Howard Florey, 1944 (University of Adelaide Library, Rare Books and Manuscripts. Donor: Dr Joan Gardner); US Navy radar scope in use during Second World War (Alamy); (back) colonies of penicillin mould (Shutterstock); the first atomic bomb test at the Trinity site, New Mexico, 16 July 1945 (Wikimedia Commons).
Printer Griffin Press

Extract from 'Thoughts in 1932' by Siegfried Sassoon appears with permission. Copyright Siegfried Sassoon by kind permission of the Estate of George Sassoon.

All reasonable efforts were taken to obtain permission to use copyright material reproduced in this book, but in some cases copyright could not be traced. The author welcomes information in this regard.

This book is printed on paper using fibre supplied from plantation or sustainably managed forests.

For Mum

Contents

Prologue		1
1	Dreaming in the city of churches	9
2	A life of the mind that matters	24
3	The three quests	45
4	Prelude to war	82
5	Hell breaks loose	102
6	Sunday, 26 May 1940	137
7	Alone in the storm	170
8	Missions to America	198
9	Aussie stirrers	214
10	The arsenal and pharmacy of democracy	234
11	The 'salvation of the Allied cause'	249
12	Of mice and men and melons	266
13	Critical mass	291
14	Smiling public men	317
Afterword		336
Acknowledgments		350
Sources		354
Bibliography		393
Picture credits		411
Index		412

Prologue

In the summer of 1941, within a few weeks of each other, two Australian scientists flew from a besieged Britain, across the grey Atlantic, bound for America's hopeful shores.

Old friends from Adelaide and now world-renowned in their fields, they were unaware of each other's top-secret mission. Each carried a small yet priceless cargo, among the most critical and consequential ever carried from the Old World to the New, the fruit of revolutionary breakthroughs crowning decades of scientific work.

They were flying at perhaps the most desperate time in the war for Britain, hoping to persuade a neutral United States to lend its seemingly inexhaustible industrial might to the cause of freedom – to take their inspired concepts and transform them in the white heat of American technology into mass products to fight and win the war against Hitler.

These two flights changed the course of the Second World War. Thanks to the imagination and persistence of these two Australians, critical inventions were developed and deployed in time to play a decisive role in the war.

Two Australians, two flights, three world-changing technologies – this is their remarkable story.

《 • 》

For the whole flight he balanced his briefcase on his knees. He would not let it out of his sight, carefully packed as it was with notebooks, copies of a journal article and mould samples. Nazi Germany, he knew, was interested in his research. He wasn't taking any chances. He had, though, been forced to place in the onboard refrigerator several clear glass vials containing an unremarkable brown powder. It had to be kept cold. Science trumped security.

Travelling from Bristol, in south-west England, down the dangerous 'neutral corridor' to Lisbon in Portugal a few days earlier, a bespectacled Australian scientist, Howard Florey, and his unassuming British colleague, Norman Heatley, boarded the Pan Am Dixie Clipper to fly to the Americas – and into history.

As the luxury seaplane took off at 10 am on 1 July 1941, Florey leaned back in his leather seat and contemplated the life and death struggle facing Britain and its few allies. The peril added to the urgency and importance of his mission. It was a race against time.

Having failed a year earlier to bring Britain to its knees in the Battle of Britain, Hitler now decided to starve it instead. In April 1941, Britain was suffering from unsustainable ship losses and all but powerless against Germany's elusive fleet of U-boats. Leading historian AJP Taylor concludes it 'was probably the moment when Great Britain came nearest to losing the war'.

A week before Florey left Oxford, three million German troops invaded the Soviet Union. It was the largest invasion in history. Britain was no longer fighting alone against Hitler, but for how long? Just five days into Operation Barbarossa, the German panzer vanguard was already one-third of the way to Moscow. Military experts were virtually unanimous: Hitler would crush Russia even faster than he did France the year before. And then? The victorious Fuhrer would focus his rage on Britain. The invasion, postponed in 1940, would be on again.

Back in 1940, Florey and his team had rubbed the spores of a miraculous mould into the very fabric of their clothes. That way, should they suddenly need to evacuate their laboratory, they would carry the secret with them, on them. But now, with no need for subterfuge, the means to alleviate much of the suffering in the world below travelled in Florey's leather suitcase and the vials in the onboard fridge.

Florey's mission, as he disembarked in New York mid-afternoon on 2 July, was to convince America's research institutions and pharmaceutical companies to mass-produce an extraordinary new drug. British industry had no capacity to manufacture it in sufficient quantities. But for all its apparent promise, selling penicillin to the Americans would not be easy. A revolution in human healing hung in the balance.

« • »

Five weeks later, on 5 August 1941, another Australian scientist made the challenging transatlantic crossing. His flight was a lot less comfortable, and he almost didn't make it.

A B-24 Liberator, a heavy bomber, was a long way from the luxury of the Pan Am Dixie Clipper. Cruising at 300 kilometres per hour above the dark waters of the North Atlantic, a Liberator normally took 16 hours to fly from Prestwick in Scotland to Gander in Newfoundland. It was direct and fast, but also freezing and noisy, making conversation onboard all but impossible.

For Mark Oliphant it didn't matter. He didn't feel like talking. He was desperately worried and couldn't sleep.

His official reason for travelling across the Atlantic was hopeful: he was carrying secret blueprints to help the Americans refine a recent British invention, microwave radar. While officially still a non-combatant, the United States was secretly gearing for war. Oliphant's technology would give the Allies an offensive edge against Nazi Germany – and other potential enemies like Imperial Japan.

Oliphant knew he would be received warmly. After all, his team at the University of Birmingham had developed the resonant cavity magnetron: an ingenious device that made microwaves. This technological breakthrough made it possible to shrink radar so it could fit in a suitcase, turning it into one of the most versatile and decisive instruments of waging war.

But Oliphant's journey was also inspired by fear.

British intelligence had discovered that the Germans were contemplating an atomic bomb and might already be building it. The consequences for humanity would be catastrophic. Oliphant had convinced the British government that a nuclear weapon could be built, but Britain did not have the resources to make it.

His second – secret – mission was to jolt the United States and its manufacturing might into building the bomb before Hitler did. The Australian scientist was not altogether comfortable with this task. His prewar research in nuclear physics had always been for peaceful purposes. Sure, the race against the Germans had to be won, but he was troubled.

In 1916, during the Great War, the renowned New Zealand physicist and Nobel laureate, Sir Ernest (later Lord) Rutherford, had given a public lecture in London. Rutherford, later a father figure and mentor to Oliphant, had pointed out that radioactivity was a tremendous source of energy, which some of his colleagues were keen to unlock. But he ended his talk with an ominous warning: 'I hope it never happens … or some bloody idiot might blow the world to bits'.

Oliphant must have wondered, 'Am I the bloody idiot?'

He almost did not get the chance to find out. The flight navigator made an error while approaching the North American coastline and turned the Liberator away from land. Lost and running out of fuel, the desperate pilot broke through the thick clouds and spied a newly built runway on a muddy island in the mouth of the St Lawrence River, not far from a little town called Pugwash. When he landed the fuel gauge read zero. Oliphant emerged shaken and exhausted. The next six weeks were to be the most hectic and important of his life.

« • »

The Second World War, that 'vast mechanized Iliad of suffering' as British writer AN Wilson put it, would sow death and misery for another four years after these two flights. The Allies eventually prevailed – this much is known. But the vital role the two Australians played in achieving that victory remains largely unrecognised.

Oliphant and Florey led teams that over a period of one hundred days in early 1940 developed the device that was critical to winning the war, conceived the powerful weapon that ended it, and produced the miracle treatment that enabled countless casualties to survive it. Their contribution, however, did not begin and end with science. Just as importantly, Florey and Oliphant were also instrumental in enlisting America's technological and industrial might for their cause. In another hundred-day burst of activity a year later, in mid-1941, their tireless lobbying and agitation across the length and breadth of the United States ensured the full potential of these breakthrough inventions would be realised.

Microwave radar, the atomic bomb and penicillin became the three most significant scientific and industrial projects of the Second World War. They also proved crucial to winning it. Without the two Australian scientists and their unheralded contribution in and out of laboratories, the course of war would have been far deadlier and more protracted.

Prologue

In initiating and then championing these three projects, Oliphant and Florey changed history. They are the two most consequential Australians of the war, and among the war's most consequential scientists. Their ideas and actions echo still, 80 years on.

1

Dreaming in the city of churches

In the late spring of 1921, two young men sat down for a drink not far from the sandstone halls of Adelaide University. It was a farewell, of sorts.

One was an athletic, young doctor with an intense gaze and slick dark hair, parted fashionably down the middle. Blessed with striking good looks as well as a sharp mind, he had won a Rhodes Scholarship and was soon to depart by ship for Oxford University.

His friend, a scientist, was a gentle giant, tall and broad-shouldered, with mousy, curly hair atop an oversized head. He was three years younger and had just graduated with a physics degree. Severely short-sighted and deaf in one ear, the bespectacled, benign boffin was a foil to the more self-assured and worldly doctor.

As different as they appeared, the young doctor, Howard Florey, and the young physicist, Mark Oliphant, had much in common. They shared the excitement and adventure of scientific research and felt the first stirrings of youthful ambition to write their names in the great book of human progress. They mused that fine spring day about their futures and their dreams, not yet old enough to know the difference. Six years later Oliphant would follow Florey to Britain, heading to Cambridge.

Britain beckoned the talented and ambitious. Australia had individual scholars of excellence, indeed renown, but it was too small a society to foster and sustain a rich academic culture. Before the age of air travel and the internet, it was too distant from the bubbling cultural and scientific ferment of places like London, Oxford, Berlin and Boston. As generations of migrants sailed from the Old World to Australia hoping for a better future, the quest for an intellectually richer life led Australia's migrants of the mind to retrace the route back to the old capitals of learning.

Britain might have seemed another world – but despite the distance it was never a foreign country. Sun-kissed Antipodeans might have been objects of some curiosity and fascination to Britons, but 'colonials' like Florey and Oliphant would never feel like strangers in a strange land. South Australians, like Canadians and New Zealanders, were British subjects all: scattered children of, as they had all been taught, the greatest empire the world had ever seen.

《 • 》

They were born three years and three kilometres apart, straddling both the turn of the century and the span of the city centre. But they were also born in different dominions: Florey in a British colony, Oliphant in the newly minted Commonwealth of Australia.

On 1 January 1901, in Adelaide's Town Hall, the state's governor, Lord Tennyson, addressed a crowded and enthusiastic audience of South Australians. The eldest son of the bearded bard of Empire, Alfred, Lord Tennyson, the governor welcomed the queen's subjects to a new nation and a new century. Like his famous father, the governor was an ardent imperialist. In his remarks he noted happily that 'the Imperial federation was no longer a mere vision', reflecting that 'the closer and more intimate the union between the different states of the Commonwealth was, the better it would be'. Assiduously stroking local egos, Lord Tennyson told assembled guests that he had declined an invitation to attend Sydney's larger celebrations as he preferred to be among them on this momentous day. The festive assembly lapped it up. 'I thought the clapping would never end', Lady Tennyson reported to her mother.

Adelaide in 1901 was the third largest city in Australia, with some 170 000 people, or about half of the state's population, calling it home. Few of the city's Indigenous inhabitants remained. The Kaurna, the first people of the Adelaide Plains, were scattered to missions and rural settlements. But Adelaide's new residents would by and large agree with Lord Tennyson; they would much rather be here than anywhere else. A city of parks and churches – 'About 64 roads to the other world', according to Mark Twain, who with 'cosmopolitan curiosity' tabulated a list of local religious sects and denominations during his 1895 visit – it retained the idealistic, if not slightly utopian, promise of its visionary founders. Spacious, clean and well run by its progressive municipal authorities, Adelaide gave off 'a general air of comfort and well-being that contrasted with the extremes of wealth and poverty found in the larger cities'. Such a setting would make for comfortable, even idyllic, childhoods.

Relative isolation which gave rise to a self-contained air was a fact of everyday life. In 1901, few Australians – and even fewer South Australians – travelled. Journeying overseas was by steamer and cost a small fortune. Communication proceeded by letter or, if you were rich, by telegraph. There was no such thing as 'aviation', just a few crazy men dreaming of flying machines, both primitive and dangerous. In 1901 motor vehicles were a rarity on the dirt and gravel thoroughfares of Adelaide – though the Floreys were soon to own the 17th car in the city and its second with a roof. For most residents the daily commute was by buggy, if not by foot. The streets smelled of horses, their straw and their manure; life and days 'perfumed … with saddle-oil, joss-stick and railway steam'.

Despite the isolation, Adelaideans never felt lost or alone. The colony, if not its Indigenous peoples, basked in the familiar glow of Pax Britannica. For Adelaide patricians and politicians as much as publicans and porters, London was a new Rome: the capital of the world. By and large they considered themselves to be fortunate: part of a great, globe-spanning community of shared heritage and values, united by their faith (if not always by sect), their monarchy and their civilising mission to the world. A far-flung family it might have been, but like every family it brought a sense of stability, comfort and belonging.

While acknowledging – and celebrating – their continuing fealty to the Mother Country, most Australians had by the end of the 19th century also come to believe in the need for continental unity and self-determination. The budding nation – as it now increasingly thought of itself – was growing fast in population and prosperity, making existing colonial arrangements unwieldly and fraught. Much better to have a national government to administer defence, foreign relations and other crucial matters, all within the framework of the Empire, of course.

A bill providing for the creation of the Commonwealth of Australia was among the last signed by Queen Victoria during her 64-year reign. The aged matriarch of Empire frailly witnessed Australia achieving nationhood on the first day of the new century and passed away 21 days later. It was the end of an era, with British power at its zenith. But the imperial tide was already ebbing, even if most remained oblivious to the first signs of decline. Despite setbacks in the war in South Africa, the map of the world on the wall of Florey's and Oliphant's classrooms still glowed a comforting crimson showing far-flung possessions. British subjects around the world remained optimistic, willing only to contemplate a future that was an extension of the past.

《 • 》

As the 20th century began, science, ubiquitous and inescapable, had emerged as a defining presence in modern history, its many fruits seen as further evidence of the greatness of the civilisation that produced them. And while British imperial progress might now have stalled, scientific and technological advances looked unstoppable.

In the year of Australia's Federation, Italian inventor Guglielmo Marconi captured the world's attention when he received the first transatlantic radio signal. The nature of radio waves and electricity became popular areas of exploration and experimentation. In Britain, physicist JJ Thomson had just discovered the electron, sparking great interest in the study of the atom, the building block of matter. Meanwhile, a young New Zealander, Ernest Rutherford, already a professor at McGill University in Montreal, was studying radioactivity, which led to his 1908 Nobel prize in chemistry.

Also in 1901, the first ever Nobel prize in Physiology or Medicine was awarded. The proud recipient was German physiologist Emil Adolf von Behring, whose discovery of serum therapy, in the words of the Swedish Academy of Sciences, put 'in the hands of the physician a victorious weapon against illness and deaths' (Von Behring developed 'anti-toxin' vaccinations against diphtheria and later tetanus).

The new century held the promise of many more such developments in humanity's oldest war.

As with imperial glory, even the most remote outposts could share in the story of scientific advancement. At the University of Adelaide, William Bragg, the professor of mathematics and physics, was already at work on 'analysis of crystal structure by means of X-ray', for which in 1915 he and his son, Lawrence, would share the Nobel prize for Physics. Aged 25 at the time, Lawrence Bragg remains to this day the youngest science laureate, and second youngest Nobel winner in any field.

The Bragg father-and-son team might have been an inspirational local success story for budding scientists like Florey and Oliphant, but they also left Adelaide, moving to Britain in 1908 to carry their groundbreaking research to its fruitful conclusion. There were limits to local achievement. Yet regardless of place, 'scientist as hero' would inspire generations to devote their lives to expanding the frontiers of knowledge as surely as explorers and warriors, those more traditional childhood idols, had once done with the frontiers of empires.

As the 20th century began, men and women in Western countries believed that science and technology would make everyone's life better: people would live longer and healthier lives, labour would be safer and easier at home and at work, everyone would be better connected and better travelled. After all, their daily existence was already markedly different from that of their parents and grandparents. They also hoped, a recent sentiment unprecedented in human history, that the lives of their children and grandchildren would be better still, change being not just possible but inevitable and inevitably for the good. If the Ancient Greeks invented happiness, it took the Enlightenment to bend the Western mind to optimism and progress. And it took science and technology to begin to deliver on this hope for the many and not just the fortunate few.

The world that awaited the arrival of Howard Florey and Mark Oliphant was brimming with excitement and on the cusp of great change.

《 • 》

Howard Walter Florey was born in a tin-roofed stone cottage in Malvern, an inner suburb of southern Adelaide, on 24 September 1898. Marcus ('Mark') Laurence Elwin Oliphant was born three years later, on 8 October 1901, in a small, rented bungalow in the inner Adelaide suburb of Kent Town, just to the east of the city. Their homes were separated by the suburb of Parkside (now including the Howard Florey Reserve) and the Adelaide Parklands, ringing the city centre in a pioneering vision of urban planning.

Thanks to the rising prosperity of their middle-class families – Joseph Florey was a successful immigrant bootmaker and businessman, and Harold Oliphant (known as 'Baron') an engineer and public servant – Howard and Mark both grew up in the comfortable and breezy enclaves of the Adelaide Hills to the east of the city. They enjoyed tranquil childhoods at the edge of the great Australian outback. Howard, the youngest after four sisters, often wandered alone, daydreaming through fields of kangaroo grass and eucalyptus forests that extended beyond the fence of the family home. Mark, the oldest of five boys, never lacked for company, though he was not as robust and physical as Howard. Memories of childhood adventures – and the fragrant odour and brilliant light of the Australian bush – were to comfort them both for a lifetime, particularly in moments of nostalgia and reflection during the long years in often wet and grey Britain.

Not surprisingly, Howard and Mark excelled academically. Both families had moved back to the city by the time of high school. 'I lived in the same suburb of Adelaide as Florey', Oliphant later recalled, 'but he went to St Peter's College, which was a snooty school, I went just to a state school [Unley High School and then Adelaide High School], but we knew one another even in those days. And we were both interested in science'. Howard arrived home from school one day, not yet a teenager, and announced that he would devote his life to scientific research. When his sister, Hilda, then studying medicine at the University of Adelaide, kidded him, 'Oh, you'd like to be a sort of Pasteur?', young Howard had no idea who that was. But Hilda's words would prove prophetic.

In turn, Mark built, apparently with his parents' approval, a 'laboratory' underneath the new family home in the suburb of Mitcham. But not all his experiments were known to further science. On one occasion, as Guy Fawkes Night approached, Mark decided to make his own fireworks. Persuading a younger brother to 'borrow' some brass doorknobs from a palatial home nearby, Mark made gunpowder which he proceeded to pour into the knobs. His brother, Keith, recalled that the subsequent explosion 'nearly blew Mitcham off the map'. This too was a powerful, if eerie, portent of his future.

Howard topped physics and chemistry in the classroom – though, curiously, not biology – and also dominated the sporting field. He captained football, tennis and cricket teams, and fiercely competitive, gave no quarter and no thought to offending others. He was a born leader. Mark's early life was more exclusively one of the mind. His love of books and learning, influenced by his mother, Beatrice, and a strong moral sense inherited from his Theosophist parents, lasted a lifetime. Oliphant much later described his parents as 'do-gooders'. While strict religious faith left him relatively early, Oliphant explained that he had a sound relationship with God, if not with all the 'fat prelates in Rome, London, New York or Sydney', and a strong social conscience. As he fought off his early shyness, he became a man of confidence, humour and bearing – qualities he would draw on in the years ahead. What came more naturally to Florey, Oliphant had to learn.

《 • 》

In 1917, Florey left St Peter's with a government bursary and headed to the University of Adelaide to study medicine. After the revolutionary discoveries of Pasteur and Lister, medical research was enjoying a renaissance. Florey yearned to be part of this new age of healing. His ardour for mastery of anatomy, biology, physiology and bacteriology was rewarded with more prizes and awards. Florey never doubted he had found his calling.

Yet, like many young Australians of his generation he suffered from guilt. In wartime Adelaide it was not uncommon for apparently able-bodied young men wearing civilian clothes to be insulted or presented with a white feather as a symbol of cowardice. Caught up in a tide of nationalism and feeling pangs of remorse, Florey twice sought to join the army that his father's boot company shod. But his parents, especially his mother, Bertha, bitterly opposed this. Florey's battles with pneumonia and his position as the only son caused him to rethink. He was further dissuaded from enlisting when the government finally issued special badges to medical students, saving them from the indignity of explanation. After a crooked accountant broke his father's business and then his health, leading to his early death in September 1918, Florey abandoned the idea of military service for good.

Oliphant was spared such dilemmas. Three years younger than Florey, he finished at Adelaide High School just as the Armistice was brokered. But although too young to enlist, he was touched and scarred by the war. Like Florey, he knew older school mates who had served and died. By the end of the war, over 60 000 Australians, including 5000 South Australians, would lie under rows of white crosses in foreign fields. When the diggers, older and hardened, returned to parade in Adelaide, Oliphant joined in the cheering. He was convinced that there would never be another big war like it. Surely, people could not be so stupid.

At university, Florey also found time for love. Ethel Hayter Reed was intelligent, ambitious and vivacious. Three years behind Florey, she was the only woman in her year at medical school (as Florey's older sister, Hilda, had been in hers). Not surprisingly, she became the centre of attention in the small world of medical undergraduates in wartime Adelaide. So began a courtship that would continue over long distance, as Florey built his career in Britain. They married in 1926.

But all that was in the future. In 1920, having excelled academically and at sport, Florey was encouraged to apply for a Rhodes Scholarship. To his great joy he was successful, neatly fulfilling the manly injunction laid down by Cecil Rhodes: the recipients were not to be mere bookworms and scholars but should display a 'fondness of and success in manly outdoor sports and qualities of manhood, truth, courage, devotion to duty, sympathy for and protection of the weak, kindliness, unselfishness and fellowship, moral force of character and of instincts to lead'. Young Howard had all that and more.

Brilliant, handsome, ambitious, driven, obsessed and idealistic, Florey said goodbye to his family and friends –

including Oliphant – and boarded the steam ship SS *Otira* at Port Adelaide on 11 December 1921, bound for England, 'on the Rhode to Magdalene [sic]' College at Oxford, as his university colleagues jokingly put it. He thought it likely that he would never return to live in Australia. In that he was right; everything else he could have hardly foreseen.

《 • 》

In 1919, Oliphant enrolled at the University of Adelaide. He was not yet 18 years old, pale and a little shy. Tall, curly haired and square shouldered, he was easily recognisable around the small inner-city campus. In later life, with a shock of white hair and the confidence befitting a highly influential Australian, he was called 'handsome' and 'distinguished-looking'. But his friends and colleagues always remember his booming laugh. He disarmed those who met him with informality and good humour. In the years ahead, when, like Florey, he needed to badger and bend countless ears to attain his ends, Oliphant would often do it with a smile.

He first thought he would be a clergyman, then a doctor. But he was inspired by Brailsford Robertson, the Professor of Physiology and Biochemistry at the university, and Oliphant decided he did not want to spend his life treating patients. He wanted to experiment, he yearned to be a scientist.

This was fortuitous when his family fell on hard times and were not able to afford the medical school fees. Already known for his practical aptitude with gadgetry and apparatus, the Physics Department offered him a cadetship with free tuition and a small income. He described his work as the laboratory 'dog's body' – but he loved it and it changed his life. He was captivated by physics and chemistry. His practical ingenuity was soon matched by intellectual capacity and he obtained a Bachelor of Science degree in 1921, followed by First Class Honours the year after.

Upon graduation, Oliphant remained at the University of Adelaide as the laboratory assistant to the head of the Physics Department, Professor Kerr Grant, who had previously taught Florey. Adelaide University in those days was still a place of studious intimacy and quiet contemplation, where 'laboratories smelt of tar and resin, lecture rooms of ink and chalk' and most staff and students knew one another, at least by sight. As the third decade of the 20th century dawned, only a few hundred rubbed shoulders at the North Terrace campus. It was a small, self-contained world.

Soon Oliphant was responsible for setting up the most complex of experiments and encouraged to initiate his own research. Unshackled, he began to publish in leading journals, including *Nature*, and soon submitted articles solely in his own name. Not yet 24, Oliphant was a rising star.

In September 1925, the tiny world of Adelaide physics was struck by a legend: a visit from the father of atomic physics and the first man to 'split' the atom, Sir Ernest Rutherford. Rutherford, a Noble laureate, was now the head of the prestigious Cavendish Laboratory at the University of Cambridge, which was emerging as a centre for physicists from around the world.

Oliphant was starstruck. As he would later recount, 'I was a nobody. I didn't even get to meet or speak to him. But his words and his personality electrified me. I determined I must find a way of working in his laboratory'. It seemed a pipedream, but ambition and confidence were biting at the self-effacing young physicist. Oliphant decided to apply for an 1851 Exhibition Scholarship that would take him to Cambridge. Backed by Kerr Grant's enthusiastic support and a growing record of publications, he was successful. He could now fulfil his dream of working with Rutherford, a man, Oliphant later reflected, 'I grew really to love'.

Oliphant met Rosa Wilbraham at the university dancing club. Smitten, he forged a wedding ring for her in the university laboratory from a gold nugget gifted by his father and they married in 1925. It was the beginning of a partnership that endured more than 60 years. Oliphant and Rosa enjoyed a happy and loving marriage. She was his muse and inspiration in a life of high achievement.

Cambridge, where the young couple arrived in October 1927, was the second great love of Oliphant's life. Here, at the heart of British science, the 26-year-old up-and-coming man of physics was now also joining his friend, Howard Florey, the 29-year-old doctor-turned-researcher, who had transferred from Oxford to Cambridge to undertake doctoral research. After six long years, the two young scientists from Adelaide were reunited on the other side of the world.

2

A life of the mind that matters

By the beginning of the 20th century, popular opinion had crowned modern science as humankind's most significant and consequential achievement. In a world lit by electricity and powered by steam, scientific work was judged the miraculous pinnacle of human creativity and imagination.

Physics had just entered its Golden Age. Human reason began to contemplate a reality it could not see and was barely able to conceive. Speculation and argument about the nature of matter – the fantastic world-within-a-world of atoms and sub-atomic particles – was not based on observation so much as insight. It was more metaphysics than physics as Galileo and Newton had known it. Rutherford first hypothesised neutrons in 1920, but it took another decade to establish their existence. So, too, with Einstein's work on relativity. Imagination grasped at possibilities – a theory or explanation – years before it could conjure up proof.

Jacob Bronowski later described the glory days of physics as 'the greatest collective work of science – no, more than that, the great collective work of art of the twentieth century … The human imagination working communally has produced no monuments to equal it, not the pyramids, not *The Iliad*, not the ballads, not the cathedrals'. For those like Oliphant, physics was not just interesting or even exciting. It was revelatory – almost a religious experience. Physics posed questions that went to the essence of everything. It was 'walking the path of God', reflected fellow physicist Isidor Rabi. 'When you're doing physics, you're wrestling with a champ … You're trying to find out how God made the world, just like Jacob wrestling with the angels'. It was transcendent and poetic; language stretched to its limits to describe the unfathomable through images and metaphors. Physics was an adventure at the edge of certainty, at the edge of comprehension; a sublime odyssey.

The science of medicine was also enjoying a revolution.

If the advent of dissection during the Renaissance went some way to improve knowledge of the human body, treatment of disease had changed little for over two thousand years. Well into the 19th century, remedies like bleeding and purging were universally applied in order to rebalance the four humours – blood, phlegm, black bile and yellow bile – fluids believed to circulate inside the body and account for one's health and disposition. In 1799, George Washington was bled to death by the best doctors of the day as they attempted to cure his sore throat.

Then, at last, in the second half of the 19th century, the Frenchman Louis Pasteur proposed that minute bacteria are responsible for many of our ills – which became known as the 'germ theory of disease'. This simple yet revolutionary insight provided the conceptual leap necessary for the development of effective sanitation, vaccines and medicines responsible for the subsequent staggering advances in human health, healing and longevity. The Englishman Joseph Lister, who first developed antiseptics, and the German Robert Koch, who identified bacteria responsible for specific infectious diseases, soon joined Pasteur in revolutionising medical science.

In 1905, William Osler, the leading teacher of medicine in the United States, wrote that more had been accomplished in improving human health in the 19th century than at any period in history. In truth, it was more like the second half of the century. But understanding the causes of many diseases, and even sometimes being able to prevent them, was one thing; curing them was another. For medical researchers in the infant sciences of bacteriology and pharmacology the journey was only beginning.

The young Howard Florey was spurred by the possibility of discovery – the rush to be the first and the rush one gets from being the first. Florey wanted to peer inside the human body and find answers. Answers to questions that had thus far challenged and stumped the great figures of medicine; answers for the ills that had forever plagued hapless humanity. If perhaps not as sublimely transcendent, the world of cells and microorganisms might be as beautiful

A life of the mind that matters

to its beholder as that of atoms and particles. It could also seem more practical and grounded, with its promise to both satisfy the curious and comfort the suffering.

《 • 》

Florey arrived in Oxford in January 1922. It was freezing, snow-covered and like nothing he had ever seen before. He immediately fell in love with it, describing his chosen college, Magdalen, as 'undoubtedly the most beautiful college here … and the High! The most beautiful street in Europe!' This was a big call for someone who had only ever lived in Adelaide, but over the centuries few have disagreed with Florey. It was the Oxford of his dreams, the place to fulfil his potential and burnish his ambition. Leaving Adelaide had been hard, but at 23, Oxford was the key to his future.

The great university seemed to be just emerging from the Middle Ages, having only recently abolished compulsory Greek and begun admitting women to most degrees – a mere half century after the University of Adelaide. Having supplied the politicians, diplomats, lawyers and administrators of Empire for generations, Oxford (and Cambridge too) was now haunted by its lost generation: nearly 20 per cent of students who served in the Great War perished in the trenches, almost double Britain's overall casualty rate (half of Oxford students volunteered in the first months of the war alone). By the time Florey arrived, undergraduates were not usually returned servicemen, but the shadows of war lingered.

Not surprisingly, elite universities became for some an escape from the crushing weight of the recent past. The languid aestheticism of Oscar Wilde made its last return, providing a glorious setting for Evelyn Waugh's novel *Brideshead Revisited*. 'Oxford, in those days', wrote Waugh, 'was still a city of aquatint ... [H]er autumnal mists, her grey springtime, and the rare glory of her summer days ... exhaled the soft airs of centuries of youth'.

But Howard Florey was no Charles Ryder and certainly not Lord Sebastian Flyte. As much as he appreciated the beauty of Magdalen (ironically, Wilde's old college), Florey found its 'Englishmen ... a queer lot ... the majority preserve a frigid silence'. Most undergraduates, he wrote to Ethel, 'annoy me excessively. The Aesthetes, who wear gilded ties and other odd things are nauseating'. The cold college and its even colder residents were a shock.

If the Englishmen (for they were nearly all men) were a disappointment, much else around Florey seemed familiar, if on a much grander scale than in Adelaide. Britain in 1922 was still the centre of Empire and world commerce, if increasingly beset by nervous shudders. The economy never quite recovered from the disruption of the Great War and spluttered throughout the 'Roaring Twenties'. Its once paramount competitiveness was progressively lost, gallingly, to defeated Germany and, more benignly, to the rising giant across the Atlantic, the United States. From cars in Detroit to celluloid dreams in Los Angeles, Americans were proving masterful at turning wants into needs and producing them cheaply for a mass market. The slow drift of global power to nations which better harnessed scientific discovery and technological innovation was becoming difficult to ignore.

Among this unsettling rumble and growing uncertainty, Oxford remained a place of refuge for the serious and the striving. Florey, in any case, was too grounded and too studious to be much distracted by either the glitz and glamour of popular culture or the more elite excesses of the 'Bright Young Things'. Florey knew Oxford was where he wanted to be. He settled into his course, determined to make his mark.

《 • 》

Florey enrolled in physiology and alongside new knowledge and experience soon acquired friends and mentors who would shape his life and career. First among them was Sir Charles Sherrington, a world-renowned scientist, Nobel laureate and, when they met, President of the Royal Society. The presidency was the most distinguished role a scientist could achieve in Britain and one that Florey himself would assume 40 years later. Eminent physiologist and neurologist, athlete, book collector and poet, Sherrington taught and guided Florey, providing him with rigorous training to become a great experimenter in his own right.

Moreover, Sherrington had worked with both Robert Koch and Joseph Lister, and was a friend of Ernest Rutherford and other legends of science. Florey was only one degree of separation from some of the most consequential men in history. It was heady stuff. Florey told his friends back in Australia, 'The atmosphere is so different here. They take an interest in you, and if you show any promise they give you encouragement'.

While driven by work and 'damnably lonely', Florey made an effort to strike up acquaintances. Most of his new friends tended to be, like him, foreigners. One was a fellow Rhodes Scholar, Keith Hancock from Victoria. Hancock was hailed as the youngest professor in the British Empire when he accepted a chair of modern history at the University of Adelaide in 1924 where he wrote the influential work *Australia* (1930). He returned to Britain in 1934 to take up the chair of modern history at the University of Birmingham where Mark Oliphant joined him in 1937. All three were involved with the founding of the Australian National University in Canberra after the Second World War. Florey's other Oxford mates, also Rhodes Scholars, were Hugh Cairns from South Australia and John Fulton from the United States, the latter also at Magdalen. Both men would become important allies and play indispensable roles in Florey's wartime quest to produce penicillin.

Despite his movie idol looks, friendships never came easily to Florey and throughout his life he rarely overcame his loneliness. He did not possess an easy personality, something that he recognised, writing to Ethel, 'I can see myself developing into a rather nasty product'. Even in his mid-20s he demanded a lot of himself and others. His initial reserve hid a blistering temper which he drew upon for the pompous, lazy and, sometimes, plain unlucky. Even as an undergraduate he did not suffer fools.

Any easy-going Aussie charm Florey possessed when he arrived at Oxford had disappeared by the time he finished his BA and, with Sherrington's backing, prepared to begin his doctorate in the field of pathology at Gonville & Caius College (universally referred to as 'Caius' – pronounced 'Keys') at the University of Cambridge. Florey had been noticed, mentored and then recommended by perhaps the greatest physiologist of his day. He reported back to Ethel, 'I couldn't possibly get a better launching into experimental medicine'.

Early in 1924, on a trip to Cambridge while scouting the idea of doctoral research, Florey stayed at Trinity College at the invitation of the then Reader in Biochemistry, Jack Haldane, another giant of 20th-century science – a pioneering geneticist and evolutionary biologist. A decorated Great War soldier, brilliant lecturer and an outspoken Marxist, Haldane was as well-known at the time as Einstein. While there Florey also met Ernest Rutherford (a year before Oliphant); JJ Thomson, who discovered the electron; and Joseph Barcroft, shortly to become the Professor of Physiology at Cambridge. Florey described his company as 'about the most brilliant scientific collection in one room'. While spending most of his career at Oxford, Florey loved Cambridge, never forgetting the place that set his future direction in scientific research. Florey moved to Cambridge in the autumn of 1924 and commenced his doctoral research under the Professor of

Pathology, Henry Dean. Dean was a shrewd judge of talent and character who was important during the war, overseeing clinical trials of penicillin in North Africa.

《 • 》

'Bring me my Chariot of fire!', cried William Blake in a poem later adapted into the British hymn 'Jerusalem'. Only five years before Florey's arrival, the fictional Ben Cross became the first person ever to complete the Trinity Great Court Run, sprinting around the college courtyard in the time it took for the clock to strike 12. This was the Cambridge portrayed in sepia fidelity by the Oscar-winning movie *Chariots of Fire* and the Cambridge that welcomed Florey; the university town of great physical beauty and romance, perhaps even more magnificent than Oxford, certainly more provincial and bucolic.

Florey would sometimes stroll along 'The Backs', a sublime walk of a kilometre or so behind the great Cambridge colleges, smoking a cigarette. A hangover of Wildean decadence lingered here too. Steven Runciman, later a famous scholar of the Crusades and Byzantium, haunted the cloisters of Trinity College 'with a parakeet perched on his heavily ringed fingers and his hair cropped in an Italianate fringe'; and future society photographer Cecil Beaton sauntered around the campus in 'an evening jacket, red shoes, black-and-white trousers, and a huge blue cravat'. 'In Cambridge', recalled Florey's Cambridge contemporary, the writer Christopher Isherwood, 'there was so much to distract you from the sordid subject of work'. But the Australian doctoral student on a small stipend was allergic to cant and could not afford to wallow in tired decadence like an English public school boy. Florey concentrated on his work. He did not waver for a lifetime.

Oxford might have been more sophisticated and august, but Cambridge was more rigorous, analytical and scientific. It had been home to Francis Bacon, Isaac Newton and Charles Darwin. The poet William Wordsworth left before graduation, driven away, he claimed, by the university's 'mathematical gloom'. For Florey it was heaven.

He commenced the doctoral research that eventually became his thesis 'Physiology and pathology of the circulation of the blood and lymph'. Sherrington had first caught Florey's promise, so now did Dean, Jack Haldane and Edgar Adrian, the physiologist who, like Florey, would go on to win the Nobel prize (with Charles Sherrington) and serve as President of the Royal Society. But while Florey revelled in their company and might occasionally even indulge them with his charm, his behaviour towards his peers did not improve.

'His drive and ambition were manifest almost from the day he arrived', recalled Sir Alan Drury, later a friend and collaborator:

> A great fire seemed to burn within him, and his many-sided character was never concealed. We could all see the power in him and wondered whether he would ever find the

right outlets for his greatness ... He could be ruthless and selfish; on the other hand, he could show kindliness, a warm humanity and, at times, sentiment and a sense of humour.

Scathing and cutting to others, he was never known to go easy on himself. 'He displayed utter integrity and he was scathing of humbug and pretence. His attitude was always – "You must take me as you find me"'. This wasn't always easy. 'To cope with him at times', Drury remembered, 'you had to do battle, to raise your voice as high as his and never let him shout you down ... but if you insisted on your right he was always, in the end, very fair'.

Florey confessed to hiding behind the Australian stereotype as a ploy or explanation for his bad behaviour: 'I could always get away with being audacious and do the outrageous thing and still be tolerated. They made allowances for the rough colonials and some people came to expect that sort of behaviour. It was always a good line of attack'.

Yet for all the rough and ready Antipodean caricature – Florey sometimes referred to himself as 'the bushranger', as did others – his professional stocks continued to rise. Before completing his doctorate, and with the backing of Professor Dean, Florey won a Rockefeller Foundation travel grant to the United States for ten months in 1925–26. The grant further enhanced his experimental skills and also put him in touch with scientists who would later prove crucial in his penicillin quest. Chief among those taken by Florey's ability and dubbing him a 'rough colonial genius' was the

Philadelphia pharmacologist Dr Alfred Newton Richards, the future wartime chairman of the Committee on Medical Research, and after the war the President of the National Academy of Sciences. The trip was the start of Florey's invaluable association with the Rockefeller Foundation.

On returning to England in May 1926, Florey briefly took up a Research Fellowship at the London Hospital where he was soon joined by Ethel. After five years and 130 letters since leaving Adelaide, Howard remained deeply attached to her. Lonely and largely solitary, Florey shared his hopes and dreams with Ethel. He could not do that with his contemporaries, their relationships too often tinged with competition. Ethel, as a friend and correspondent, was a source of strength. He wooed her relentlessly, despite her hesitation. They married on 19 October at Trinity Church in Paddington. Their first child, Paquita, was born in 1929, followed by Charles (named after Charles Sherrington) in 1934.

Florey's romantic fantasy of an intelligent but submissive, adoring but charming companion was soon shattered. Ethel could not live up to the 'false image' that he had conjured in his own mind. Two strong personalities, intelligent and ambitious, they were both frustrated by their circumstances. To make matters worse, Ethel suffered from life-long poor health and progressive loss of hearing. Florey would increasingly escape into the laboratory; in turn, 'his work … became a barrier rather than a bond', wrote a colleague and biographer. The marriage was not a happy one.

In 1927 Florey had a big year: he returned to Cambridge from London, was elected a Fellow of Caius College, secured a lectureship in Special Pathology, and received his doctorate. He spent the next five years at Cambridge teaching and honing his increasingly renowned skills of experimentation and observation. In that time, he also met and recruited James Kent as his laboratory assistant. Kent would stay with him for much of Florey's professional life. Florey enjoyed five happy years in Cambridge before departing in 1932, somewhat surprisingly, to the University of Sheffield, a full professor at age 33.

« • »

The same year, 1927, was also memorable for Mark Oliphant, who arrived at Cambridge to work towards his doctorate under Ernest Rutherford at the Cavendish Laboratory. He joined Florey for five years in what they referred to as 'the colonial contingent'.

Oliphant's dream came true courtesy of a £250-a-year, three-year 1851 Exhibition Scholarship, awarded to approximately eight 'young scientists or engineers of exceptional promise'. While less known than the Rhodes Scholarship, the UK-based stipend nevertheless boasted many eminent awardees, including several of Oliphant's new Cavendish colleagues: Rutherford himself, Ernest Walton, John Cockcroft, James Chadwick, and his Australian contemporary Harrie Massey.

The Cambridge that welcomed Oliphant had changed in the three years since Florey first arrived. While aesthetes like Runciman and Beaton could still be found clowning in the cloisters, there was new-found seriousness in the air: less posing, more politics. The rise of fascism and communism across Europe put paid to childish things. So did the Great Strike of 1926, which briefly took Britain to the brink of what many feared and some welcomed as revolution and class war. 'The Germans state openly that they're going to have another slap at France and won't wait a century to do it', noted Florey on his tour of Germany in 1923. 'Germans, Austrians, Italians, Czechs, Hungarians – all fussing about their national honour, damn them. I've come to the conclusion that it's the menace of the future', he observed with grim foreboding.

Rising tension abroad slipped a discordant note into the Jazz Age and brought some new and unwelcome attention to Cambridge. In the early 1920s, the Cavendish Laboratory became an important target of Soviet scientific espionage on account of its groundbreaking work into radioactivity, X-ray crystallography and nuclear physics. There is no conclusive evidence that the Soviets managed to extract any useful secrets from the Cavendish (Oliphant had just missed the future father of the Soviet hydrogen bomb and a suspected agent, Yuri Khariton, who had a two-year stint at the laboratory), but scientific data was not the only thing the Moscow spymasters were after. Andrew Boyle, a historian of the Cold War espionage, notes, 'Nearly two-thirds of prominent British Communists or Party sympathisers during the twenties had studied at Oxford or Cambridge'.

'Oxbridge' provided a murky but fertile pool for recruitment of spies and 'agents of influence', including Kim Philby, Anthony Blunt, Guy Burgess and other members and associates of the infamous 'Cambridge Ring' that wreaked havoc with British and American security during and after the Second World War.

Oliphant did not move in these circles, however sympathetic he might have been towards progressive ideas of the day. His interest in physics was all-consuming and precluded extra-curricular pursuits. It's unlikely Oliphant even suspected his new home was one of the Soviet Union's most important espionage targets in Britain. The spies (some of whom he knew) took another 15 years to succeed. But when they did, they stole secrets that grew from the startling insight of Oliphant's research team at Birmingham.

《 • 》

Florey was vigorous in revitalising the Sheffield laboratories and got along well with his blunt Yorkshire colleagues. He was now focused on the behaviour of tissue, both healthy and diseased. Florey began examining antibacterial enzymes like lysozyme, found in human secretions such as tears and already much studied by the Scottish physician Alexander Fleming. He was tilting his mind towards naturally occurring antibacterial agents, unwilling to follow the path of the German Paul Ehrlich, Nobel laureate and vanquisher of syphilis. The promising chemical compounds Ehrlich had found proved as toxic to the human body as they were to bacteria, leading into a disappointing scientific cul-de-sac.

In Sheffield, Florey recognised he would need to collaborate with an experienced chemist in order to purify naturally occurring antibacterial agents and identity their active components. The idea of a multi-disciplinary team to tackle the problem of bacterial infection stayed with him, even though he was unable to attract sufficient funding at the time.

He didn't have to wait long. The Professor of Pathology at Oxford, Georges Dreyer, died suddenly in August 1934, and Florey was chosen to replace him. He returned to his British alma mater in the spring of 1935 to head the Sir William Dunn School of Pathology, and stayed for 27 years. It was a stroke of good fortune. Oxford gave Florey the standing to raise money and recruit new research workers. Before long he brought together a group of scientists with different skills and aptitudes who would revolutionise human healing.

« • »

A few days short of his 26th birthday in October 1927, Oliphant first made his way to the Cavendish Laboratory. It was the beginning of the academic year (known at Cambridge as Michaelmas Term) and he was about to meet the man whose guest lecture at Adelaide University two years earlier had inspired his awe. Trudging through cobbled streets and light rain, doing his best to avoid the perennial Cambridge hazard of careening student cyclists, Oliphant soon stood staring at the imposing high-arched sandstone entrance to the fabled Cavendish Laboratory, 'by far, the greatest physical laboratory in the world'.

If the façade matched expectations, the interior belied the revolution in nuclear physics unfolding within. It was a 'decrepit old building', recalled Oliphant; 'the wooden floors, dirty windows', and the floor scrubbed once a year. Even the equipment looked basic, a 'string and sealing wax' operation, all with a slightly amateurish air about it. Indeed, the annual budget for equipment and material when Oliphant first arrived was £2000. As Rutherford famously remarked, 'We haven't the money so we have to think'. And think they did, miraculously uncovering the secrets of the universe's building blocks among the weathered bricks of the Cavendish.

Oliphant navigated past the entrance jammed with bicycles, and up to the top of a cramped staircase. It was dark inside, 'a watery sun ... scarcely penetrated into the room'. The small space was 'littered with books and papers, the desk cluttered in a manner which I had been taught at school indicated an untidy and inefficient mind'. There, behind the messy desk, Oliphant remembered, 'I was received genially by a large, rather florid man, with thinning fair hair and a large moustache, who reminded me forcibly of the keeper of the general store and post office in a little village in the hills behind Adelaide'.

This was Sir Ernest Rutherford (from 1931 Lord Rutherford), one of the greatest scientists in history. Yet, he was neither pompous nor condescending, possessing a boyish enthusiasm that attracted the most gifted to his lab and the most cynical to his cause. Rutherford and Oliphant took to each other immediately, the start of a close friendship that lasted until Rutherford's death in 1937.

At this first meeting, Oliphant raised his proposed research topic: the bombardment of metals with positive ions, positively charged molecules. Rutherford 'spluttered a little as he talked, from time to time holding a match to a pipe which produced smoke and ash like a volcano', but quickly agreed. In just over two years Oliphant submitted his doctoral thesis, with Rutherford serving as one of the oral examiners.

Oliphant spent ten years at the Cavendish, the happiest time of his professional life, though punctuated by personal tragedy when his 3-year-old son, Geoffrey, died of meningitis – a disease that would become treatable with penicillin only a decade later. Oliphant escaped into research and, after a stint with gas discharges, he worked directly with Rutherford on the Cavendish's specialty, the artificial disintegration of the nucleus of atoms.

The combination of Rutherford's insight and inspired creativity – 'I feel it in my water!' he would say – and Oliphant's brilliance at conceiving and building apparatus to test deductions forged a supreme team in experimental physics. 'It was', says Oliphant, 'absolute heaven to be working with him'. As it was for his colleagues; six Nobel prizes are directly traceable to their research in 1932, often described as an *annus mirabilis*, the year of miracles in the history of physics. Oliphant was unlucky not to have claimed the seventh. Curiously, his scientific breakthrough that year generated one of the very few moments of tension between himself and Rutherford in all the years of their otherwise cordial, almost paternal, relationship.

Rutherford had told Oliphant on many occasions that he was against wasting time and energy on trying to generate power from a nuclear reaction. He staunchly believed that far more energy would be spent initiating the reaction than the reaction itself might create. Oliphant was not so sure. While Rutherford was away on a lecture tour of South Africa, he conducted experiments, using 'heavy water', showing that when two lighter hydrogen atoms are 'fused' to form a heavier atom, energy is released in the process. It is, as Oliphant subsequently deduced, the source of energy in our sun and in other stars.

Rutherford was technically right; Oliphant failed to produce more energy than he used in the experiment. But the reaction he achieved was to become the principle behind massive thermonuclear weapons – hydrogen bombs. Nuclear fusion remains to this day the holy grail of energy research: virtually limitless power with no waste product. The real difficulty remains producing it at decidedly non-sunny temperatures, in so-called 'cold' fusion. Oliphant was later nominated for the Nobel prize but ultimately overlooked.

But that was a future source of disappointment; his more immediate problem was Rutherford, who flew into a rage when he learned of the fusion experiment and delivered Oliphant a stinging rebuke for misemploying the laboratory's meagre resources on quixotic quests. 'He felt', Oliphant later recalled, 'that practical uses of his beloved nuclear physics would spoil a wonderful branch of pure scientific endeavour'. Rutherford soon calmed down, but Oliphant never forgot. He did not forever want to seek permission to conduct experiments. For the first time, his mind wandered beyond the cosy confines of the Cavendish.

It was, perhaps, the beginning of the end. The great Cavendish team started to splinter, with James Chadwick departing for Liverpool in 1935 and Charles Ellis to King's College London in 1936. Oliphant was promoted to Assistant Director of Research and Rutherford's deputy for experimental work, but now he wanted 'to run his own show'.

To Rutherford's disappointment, even distress, Oliphant left for a university where he could pursue new research with purpose-built equipment, rather than the stingy approach at the Cavendish. In the end, Rutherford agreed and supported Oliphant for the Poynting Chair and Head of the Department of Physics at the University of Birmingham, but asked him to stay on until September 1937. Oliphant gladly agreed. Rutherford also proposed Oliphant for election to the Royal Society, which, to Oliphant's delight, if not surprise, was successful.

As if marking the end of an era, Rutherford died a month later. His replacement was the Adelaide-born Lawrence Bragg, who moved the Cavendish away from a concentration on sub-atomic physics. The Golden Age was over. The next big breakthrough came from elsewhere, and almost by accident. But those halcyon days of pure research prepared Oliphant to recognise the startling implications of 'back of an envelope' calculations his Birmingham team members would make in the shadow of Nazi aggression.

3

The three quests

This book is a story of three scientific quests, which in turn became military quests. It is a story of how war – and the fear of war – accelerated scientific developments that might have otherwise taken years, if not decades, to unfold, and quite possibly along very different trajectories. And, most importantly, it is a story of two ingenious, insistent and indefatigable Australians, who were able to not only progress science but also press for its timely weaponisation, all with far-reaching consequences for the course of war and subsequent peace.

These three military quests emerged with the urgency of the Second World War, but the science behind them stretched back decades, if not centuries and millennia before. It is not possible to fully appreciate the critical role that Florey and Oliphant played in the course of the Second World War without understanding their place in the long march of scientific inquiry and progress. 'If I have seen further', Isaac Newton wrote to his scholarly rival Robert Hooke, 'it is by standing on the shoulders of giants'. Like all scientists before and after, the two Australians had to make their own climb to the summit. What their clear gaze could glimpse from their new vista – what others before them failed to see – would change war and peace.

Seeing from afar: radar

He looked like a vicar and was called 'Honest Stan'. Stanley Baldwin had served as prime minister of the United Kingdom twice already and would soon begin a third term. Now, as Lord President of the Privy Council, he was the final speaker in the House of Commons debate listed as 'International Affairs'. The topic might have been better described as 'Disarmament'.

Just before 10.30 pm on 10 November 1932, following more than six hours of contributions from his fellow parliamentarians, the tall and slightly stooped Baldwin rose from the emerald benches, smoothed his slicked dark hair, and spoke of fear and war:

> There is no greater cause of that fear than the fear of the air ... I think it is well also for the man in the street to realise that there is no power on earth that can protect him from being bombed. Whatever people may tell him, the bomber will always get through.

Baldwin was fixated on 'this terror of the air', particularly the prospect of 'bombing with gas'. So moved was he by this premonition, he believed 'it might be a good thing for this world … if man had never learned to fly'. That ship having sailed, or rather the plane having flown, he now proclaimed to grunts of approval from the government benches that 'Airforces ought all to be abolished'.

Baldwin believed that any future war would be won by swiftly destroying an enemy's military and industrial capacity from the air and bombing civilians into submission: 'The only defence is offence, which means that you will have to kill more women and children more quickly than the enemy if you want to save yourselves'.

Stanley Baldwin was leader of the Conservative Party and the largest political figure in Britain's interwar years. Winston Churchill, never a friend, thought Baldwin the most formidable politician of his day. He was no fool and no idealist. His speech in the House of Commons, in anticipation of Armistice Day the following morning, was not the solitary bleating of a defeatist. Baldwin spoke for a nation beleaguered by the Great Depression while still traumatised by the memory of the Great War. Developing technologies that turned the last war so deadly – airplanes, submarines, poison gas – made the prospect of the next one too terrible to contemplate. If the First World War was 'the war to end all wars', the next would be the war to end all civilisation. For Baldwin, disarmament was the only rational option.

Baldwin was not alone in his fears. And it wasn't just the usual suspects: the pacifists, the Labour Party, socialists and writers such as Vera Brittain, author of the haunting elegy to a lost generation, *Testament of Youth*, who cried, 'I fear War more than Fascism'. Even King George V, not exactly a wild-eyed progressive, threatened to join demonstrators shouting for peace in Trafalgar Square. Divided on so much else, radicals and conservatives, aristocrats and union leaders, the elites and the masses were all united on the need to escape the ever more deadly cycle of war.

Unsurprisingly, the Air Marshals did not agree with Baldwin on the need to abolish the Royal Air Force (RAF) for the sake of world peace. But they did share much of his analysis of the problem. Trying to stop enemy bombers is futile. The only defence is offence. You need to hit the enemy first and hit them hard. Destroy their airpower on the ground, and if necessary, destroy their military targets, factories – and cities. The very threat of such a 'knockout blow' will hopefully deter potential invaders. The RAF saw their role not as the defenders of home skies but as an instrument of a quick and decisive attack, ideally a pre-emptive one. In a world where the bomber will always get through, better it be a British one first.

In the early 1930s, out of power and out of favour, Winston Churchill believed in neither disarmament nor fatalism. A man out of touch with popular opinion, these were his wilderness years. With a belligerence eclipsed only by self-belief, and buoyed by a boyish faith in new technologies, Churchill insisted Britain must not be captive to defeatism. He argued that just as science had created a new and deadlier sword in the sky, it would – it must – also furnish a new and stronger shield on the ground.

But what would that shield be? And how did this great imperial power, which had entered the First World War with a boastful sense of invulnerability and invincibility, in a few short years come to look at the sky with such dread as to fear its own destruction?

《 • 》

The sense of dread identified by Baldwin was amplified through popular culture. Earnest authors prophesied more and worse destruction to come, often from the air, through books with titles such as *Can Europe Recover*, *At the End of the World* and *That Next War*. Science-fiction pioneer Olaf Stapledon, in his 1930 classic *Last and First Men* imagined Europe rendered uninhabitable by poison gas delivered by fleets of bombers. Cinema augmented the gloom with sight and sound. Alexander Kordas' popular 1936 movie *Things to Come*, based on HG Wells' 1933 novel, foretold of war beginning at Christmas 1940 with the bombing of London, thinly disguised as Everytown. Comic-strip characters undertook their heroics in an atmosphere of futurism and foreboding. Buck Rogers and Flash Gordon in comics, and then on the silver screen, marshalled the weapons of science fiction, including rockets and death rays, to defeat otherworldly foe, often from Mars.

Closer to home, though not necessarily reality, various cranks and oddballs were claiming that they had a death ray that could shoot an electric discharge, potentially knocking out airplanes or at least their pilots. Fanciful as it may now seem, there was 'a certain openness among the public, and, indeed, among military authorities, to the possibility of death rays and mystery rays'. There were several contenders, or perhaps pretenders, for first conceiving the death ray, including notable inventors Nikola Tesla and Guglielmo Marconi. In Germany, a certain Herr Wulle, 'chief of the militarists', informed the Reichstag that his government possessed a device that will 'bring down airplanes, stop tank engines, and "spread a curtain of death"'.

Perhaps the most colourful conman of Britain's interwar years was Harry Grindell Matthews, part mystic, part mechanic, who for a while convinced the public and the press, if not the government, that he was an inventor worthy of serious consideration. Try as they might, the authorities could never persuade Matthews to demonstrate his death ray under controlled conditions. The Ministry of Air offered a standing reward of £1000 for anyone who devised a ray that could kill a sheep at 100 yards, but no one came forward. Both the sheep and the government's money were safe.

Not so the British people. Aviation technology was progressing in leaps and bounds. With vast improvements in speed, range, payload capacity, and ability to manoeuvre, a new generation of bombers made the old German Gothas that had terrorised London in 1917 seem like mosquitos.

'Poor panic-stricken hordes will hear that hum', wrote the Great War poet Siegfried Sassoon of the warplane engines the same year as Baldwin's speech, 'And Fear will be synonymous with Flight'. The airplane's deadly potential as weapons of mass and indiscriminate destruction became all too apparent in 1935, when the Italian air force dropped mustard gas on defenceless Abyssinian villages in a one-sided struggle to conquer Africa's last independent kingdom.

Two years later, in 1937, sleek new aircraft, the apex of German engineering, were on display closer to home. In a few short, furious minutes, waves of Heinkel and Juncker bombers of the Luftwaffe's Condor Legion helped General Franco obliterate the Basque town of Guernica. This Spanish Civil War horror inspired Pablo Picasso's most famous work and terrified international audiences with images of never before seen urban devastation. It was a shocking confirmation of what many had come to believe over the previous two decades: the skies had now become a battlefield – the last of nature's domains to succumb to humanity's violent urges and potentially the deadliest. As it did over Guernica, so potentially over London and everywhere else, the bomber could – and would – always get through.

《 • 》

Abyssinia and Spain were not isolated horrors but ominous signposts to a bleaker future. With the end of Jazz Age hoopla, onset of the Great Depression, and rise of vibrant and seemingly successful totalitarian and authoritarian regimes, democracy and its liberal institutions appeared exhausted and spent. Europe's new red and black dictators were not pacifists and disarmers. They would harness science and technology to serve their resentments and belligerent ambitions. Many now began to fear that the French marshal, Ferdinand Foch, was eerily prescient in 1919 when he predicted Europe's future as 'not peace' but 'an armistice for 20 years'. If accurate, this did not leave much time.

In the summer of 1934, the RAF engaged in exercises to determine their preparedness for war. It was a disaster. Britain was virtually defenceless against air assault. Both the Air Ministry and the Houses of Parliament were quickly destroyed in mock attacks. The bombers, mock ones for now, were getting through. This sombre realisation at last concentrated minds.

The alarm raised, it took a disciplined civil servant to marshal the facts. AP 'Jimmy' Rowe studied all 53 of the Air Ministry's files on air defence and was not reassured. So, Rowe did what civil servants do – he wrote a memo to his superior; in his case, the Director of Scientific Research at the Air Ministry, HE Wimperis. Foregoing Whitehall nuance, Rowe was blunt: 'unless science can find some way to come to the rescue, any war within the next ten years is bound to be lost'.

Harry Wimperis was a mover and shaker and he was now shaken by Rowe's damning assessment. He quickly gained permission to set up a committee to 'consider how far recent advances in scientific and technical knowledge can be used to strengthen the present methods of defence against hostile aircraft'. The first meeting of the Committee for the Scientific Survey of Air Defence, later named after its chairman, scientist and mandarin Henry (later Sir Henry) Tizard, took place in late January 1935. With no time to waste, Wimperis had already consulted the effusive and voluble Scot Robert Alexander Watson-Watt, Superintendent of the Radio Department of the National Physical Laboratory, as to the possibility of a 'death ray' to knock out an airplane or kill a pilot.

Watson-Watt in turn sought advice from his colleague, Arnold F 'Skip' Wilkins, a leading researcher on electromagnetic waves. Watson-Watt posed a hypothetical to Wilkins. 'Suppose, just suppose', he said, 'that you had eight pints of water, 1km above the ground. And suppose that water was at 98F, and you wanted to heat it to 105F. How much radio frequency power would you require, from a distance of 5km?'

Wilkins smiled. He knew what the Scot was driving at. An adult male possesses about 8 pints (or over 4.5 litres) of blood at 98 degrees Farenheit (36.6 degrees Celsius) and to raise the temperature to 105 degrees Farenheit (40.5 degrees Celsius) would kill or knock him unconscious. Wilkins quickly calculated that a death ray was impossible; you could not generate enough energy in electromagnetic waves to kill people.

But Wilkins had another idea. He knew of experiments in Britain and overseas using radio waves to detect metal objects. He made a suggestion to Watson-Watt: what if you send out radio waves, bounce them off an airplane and figure out where it is? Would that be useful for air defence? Perhaps we should give that a go?

It certainly seemed a good idea to Watson-Watt who quickly penned a memo to the Tizard Committee, killing off the idea of the 'death ray' but concluding with hope that now 'attention is being turned to the still difficult, but less unpromising, problem of radio-detection as opposed to radio-destruction'. By late February 1935, focus was turning to what Watson-Watt described in his subsequent memo to the committee as the 'Detection and location of aircraft by radio methods'. Or, as we term those methods today – radar.

《 • 》

Locating the enemy is the first task of war-making. Radar is essentially seeing with radio waves, and its invention was the discovery of a new way to see. The ability to locate enemy on land, at sea or in the air, whether in darkness, in fog, or at great distance, changed warfare perhaps more profoundly than any invention since the industrialisation of combat.

While serious and competitive radar research began in the 1930s, the basic idea had its origins 50 years earlier. The most significant physicist of the 19th century, James Clerk Maxwell, on whose Scottish shoulders Einstein placed his own revolutionary work, theorised in his 1873 landmark *Treatise on Electricity and Magnetism* that both light and radio waves were electromagnetic waves governed by the same fundamental laws. This led to the conclusion that just like light waves, radio waves can be reflected from metallic objects. In a series of experiments in the late 1880s, German physicist Heinrich Hertz proved Maxwell's thesis correct.

By the turn of the century, the young Italian radio pioneer Guglielmo Marconi had established the possibilities of long-distance radio transmission, revolutionising communications and sparking further scientific interest in the properties and uses of radio waves. At about the same time in the United States, Nikola Tesla conceived of using 'electrical waves' to detect the presence and movement of distant objects, including submarines.

In 1904 a relatively unknown German engineer, Christian Hulsmeyer, performed the first practical demonstration of radio detection. A beam of radio waves from his transmitter bounced off a barge on the Rhine and back to his receiver, confirming the presence of the reflecting object. Hulsmeyer called his device a 'telemobiloscope'. But the concept of 'seeing' with reflected radio waves lingered in scientific limbo until the early 1920s, when it was picked up by Marconi, whose scientific star power brought attention back to the possibilities of radio detection.

From then on, radar was developed more or less simultaneously in several countries including Britain, the United States, France, Germany, Japan, Italy, the Netherlands and the Soviet Union. Recognising the military potential of the technology, each conducted their research separately and in great secrecy, lest scientific openness benefit current or future enemies. Britain stood out from the pack in the high priority and the degree of government support accorded to the radar project. This support was to pay big dividends.

《 • 》

Once Watson-Watt succeeded in shaking typically languid officialdom out of complacency, events moved rapidly. The Air Ministry requested a trial in early 1935 to establish whether the reflection of radio waves would in fact work to detect distant aircraft. Less than a month later, on 25 February, in what became famously known as 'The Daventry Experiment', a Heyford bomber was flown between the BBC's short-wave radio transmitter in Daventry and a receiver site. The oscilloscope to view the signal was in the back of a converted ambulance. Watson-Watt and Skip Wilkins gathered around the instruments, excited but apprehensive. But smiles soon formed as the oscilloscope's glowing vertical green line clearly indicated the plane's presence.

The implications of the Daventry experiment were clear: aircraft could be detected by radio waves. But detection was not in itself enough. The term 'radar', an acronym for 'Radio Detection and Ranging', was not coined until the early 1940s, but its pioneers knew from the start that to be useful the technology would also require the ability to 'range' what it had detected, that is, to accurately ascertain the aircraft's distance from the radar device, its altitude and its direction.

The first radar was primitive but promising. If you can detect or see aircraft coming, which early units did, and range or determine where that aircraft is going, which they hoped to do in the near future, you might be then able to direct intercepting fighter aircraft to the right spot and stop the bombers from getting through.

Watson-Watt and his team were racing against time. As German aggression on the Continent mounted, their deadlines tightened. It was not plain sailing. Some demonstrations, including one in September 1936 at the new home of radar research, Bawdsey Manor in Suffolk, about 120 kilometres north-east of London, ended in humiliating failure. A host of dignitaries, including the new commander-in-chief of Fighter Command, Air Marshal Sir Hugh Dowding, were not impressed. Watson-Watt told his men that they only had weeks to fix the problem or the entire project might be scrapped.

Fix it they did, as they were to solve many other challenges down the track. But Watson-Watt and his team did not engineer the most versatile or well-crafted equipment. As speed was of the essence, they relied on tried and tested technology to build a reliable, if sometimes clumsy, system that could be put into place and work immediately. Watson-Watt, an ever-pragmatic engineer, famously adopted the slogan 'Give them the third best to go on with, the second best comes too late, the best never comes'. He later named it the 'cult of the imperfect'.

Contemporary German radar was more technically advanced, but Nazi military hierarchy failed to recognise its full potential. When Ernst Udet, Great War fighter ace and the Luftwaffe's director of research and development, was informed of the progress German scientists had made, he breezily dismissed their efforts: 'If you introduce that thing you'll take all the fun out of flying'. Sure, radar might be useful for air defence, but the Nazi military doctrine emphasised surprise attack and swift victory. German research was transferred to the navy, where it 'languished' and radar was wasted as a mere reconnaissance aid.

What the British might have lacked in technical finesse, they more than made up for in their ability to operationalise the technology. As Churchill remarked after the war, the strength of British radar lay in 'the extent to which we had turned our discoveries to practical effect, and woven all into our general air defence system'. Unlike their counterparts elsewhere, including in Germany, British boffins understood what radar could do best, they made it fit for that purpose, and they successfully linked it with other technologies to get the most out of it.

By the outbreak of war in September 1939, Watson-Watt had supervised the construction of a radar network called Chain Home: 21 operational early warning radar stations stretching from Southampton, 110 kilometres south-west of London, to Tyne, about 450 kilometres north of London. The Chain Home network was supplemented by 30 'Low' or smaller stations built on high ground or on the coast. Chain Home could detect an aircraft approaching from about 200 kilometres away, and so still deep in Belgian or French airspace.

This mesh of radar was an 'invisible bastion' against hostile aircraft. Connected by the telephone to the 'Filter Room' at Fighter Command in Bentley Priory, on the outskirts of London, radar stations would alert the Command of the strength and the direction of incoming enemy aircraft. The Filter Room operators, who plotted all aircraft movements over the British Isles on giant maps, would phone the airfield nearest to the enemy's flight path so they could 'scramble' its fighters into position in a matter of minutes. It was radar but it was so much more. British air defences were effective precisely because the Chain Home network – and a true network it was – was much greater than the sum of its parts.

Yet the science on which the system was based had severe limitations. Long-wave radio waves used by Chain Home radar required large antennae to work, making the technology suitable only for static installations. These earthbound radar installations were good for detecting and ranging squadrons of enemy aircraft incoming across the English Channel but little more.

Watson-Watt had drawn official attention to the need for radio waves of very short length – microwaves – as early as 1935. Microwaves had enormous advantages over long waves: greater immunity to jamming, greater range and directional accuracy, fewer misleading reflections off the ground, and the ability to discriminate between small closely located targets. But most importantly, a microwave generator also required a much smaller antenna, meaning that radar sets could get off the ground and be fitted to aircraft and smaller naval vessels.

Such portable radar would be able to detect enemy aircraft, surface ships and submarines a long way from land, aid long-distance navigation, and direct the aim of individual anti-aircraft batteries and naval guns. Not just a shield anymore, microwave radar might allow the swordsman to range wider, see better and strike more surely.

As war loomed in late 1938, British top brass were well aware of how critical microwave technology might soon become. Research was urgently commissioned by the Admiralty at Britain's top research institutions, the University of Oxford's Clarendon Laboratory and the University of Bristol. A previously moribund physics research laboratory at the University of Birmingham was also asked to help. To everyone's surprise, the Birmingham turtle would humble the Oxford and Bristol hares in the race to build a device for use in portable battle-fit radar. Just as surprising, Birmingham's winning research effort would be headed by a relative novice and outsider, a 37-year-old Australian physics professor named Mark Oliphant.

« • »

Even though its work was heavily guarded, the renown of Bawdsey Manor soon spread. Watson-Watt and his team played host to a stream of dignitaries. As if to duchess the success of the radar team, a not so humble backbencher appeared at the manor on 20 June 1939. Winston Churchill had long taken an interest in radar. Though not in possession of a top security clearance, Churchill knew through his own informal intelligence network about Watson-Watt's first radar experiment at Daventry within 24 hours of its success in February 1935. Appointed by Baldwin to the Tizard Committee four months later, Churchill kept abreast of all subsequent developments.

Treated to a successful demonstration, a gleeful Churchill declined to celebrate with but a cup of tea, demanding instead 'a brandy – a big one'. Consuming his drink with 'evident relish' and never one to miss the opportunity to perform, Churchill rose to his feet. Alive with the liquor and wielding a fresh cigar, he addressed the bemused scientists and awed airmen of Bawdsey Manor. Living up to the legend, he delivered an inspirational 15-minute off-the-cuff speech before donning his black homburg and, with no little ceremony, speeding back to London.

Just ten weeks later, Germany invaded Poland, and on 3 September 1939, Britain declared war on Germany. The same day, ending his decade in the political wilderness, Churchill was appointed First Lord of the Admiralty. This enthusiast and sponsor of new technologies from the pioneering Great War days of the tank onward was now instrumental to Britain's war effort. Nine months later, upon becoming prime minister, he would be at its very centre. Oliphant and his team at Birmingham University would find a willing ally in perhaps the most scientifically literate politician to thus far occupy No. 10 Downing Street.

Searching for a cure: penicillin

Alexander Fleming was no stranger to infection. A short and sturdy son of an Ayrshire farmer who inherited his father's bright blue eyes but not his money – the family property lost to bad decisions and bad luck – Fleming had followed his older brother to become a doctor, studying at St Mary's Hospital Medical School in London before his appointment as assistant bacteriologist within St Mary's research department. Fleming was lecturing at the school in 1914 when called to arms.

Captain Fleming worked at the British Army General Hospital no. 13 in Boulogne, incongruously housed in a former beachside casino on the Côte d'Opale. But a few years into the Great War there was nothing plush or enticing about the once grand building. In this casino-cum-hospital the odds were stacked against the patients, lying and dying on soiled camp beds where roulette and baccarat tables once stood.

While much was known by this time about the cause of infection and its dangerous course, effective treatment had eluded practitioners. In this regard the learned physicians of Belle Époque Britain had advanced little from the witch doctors and medicine men of antiquity. Fleming, having worked at St Mary's under Sir Almroth Wright, a pioneering bacteriologist and immunologist, was fascinated by this grim problem. During the war he recorded the daily suffering of his wounded patients. Fleming noted in terse and untidy medical shorthand that nearly all wounds were infected. As often as not this meant death, bacteria succeeding where bullets had not. There was little he, or anyone else, could do about this deadly inevitability.

Experimenting in his makeshift laboratory not far from the frontlines, Fleming showed that antiseptics applied on the surface of the wound could not reach into the crevices and recesses of deeply infected tissue. Even worse, the antiseptic dressings compromised the body's own defences, making infected wounds even more deadly. Fleming recommended that wounds be washed and sutured and then kept clean and dry, but his advice was rarely followed.

《 • 》

To stay alive and remain healthy is humankind's oldest, most formidable and universal struggle. It is a battle fought anew every day, and for us it is hard to imagine it ever being fought without the aid of antibiotics. A world without antibiotics, however, is only 80 years ago; the world of our grandparents' or even parents' youth, many of whom owe their lives to these magic drugs.

For much of human history, life expectancy hovered between 30 and 40 years of age. This average hid a more troubling reality: that one in two children would never grow to adulthood. The lucky ones who did, after lives typically punctuated by sickness and suffering, died in middle age. To grow old was rare; envied by some, pitied by others. Our ancestors did not particularly fear cancer, heart disease or dementia; they did not live long enough to develop diseases of old age. But they were ever vigilant of infections. They did not know the cause of diseases like cholera, syphilis or bubonic plague, or why wounds and cuts became infected, but they recognised the symptoms and the consequences: swelling, pain and smell, followed frequently by death.

Around half of the ten million soldiers who died during the First World War perished not from bullets or bombs but from infection. The French and Flemish farmlands of the Western Front, rich with manure and fertiliser – and decomposing bodies – were a septic nightmare. Deadly bacteria from cultivated pastures and muddy pools soaked into uniforms and were then punched into flesh by bullets and shrapnel.

The world was still reeling after four years of war when the influenza pandemic commonly dubbed the 'Spanish flu' struck. Ironically, there is some evidence that the virus originated from Etaples in northern France, a large training and hospital centre for Allied troops, just 35 kilometres south of where Alexander Fleming was stationed in Boulogne. The virus is typically held responsible for 20 to 50 million deaths, though some estimates run as high as 100 million victims, or about 3 per cent of the world's population. Most of the deaths were from bacterial pneumonia, a common secondary infection associated with influenza that today is treatable with antibiotics.

《 • 》

The three quests

Well into the second decade of the 20th century, infected wounds and infectious diseases killed millions. But armed with insight from the 'germ theory of disease', developed over the previous 50 years, scientists sought to kill harmful bacteria inside the human body. Some looked to chemical compounds, which generally proved too toxic for internal use. Others focused on antibacterial substances of biological origin which they thought had a better chance of being non-toxic and effective in all conditions. But it still took unimaginable luck and a stroke of detective genius to finally capture and recognise the magical properties of the breakthrough substance, a humble mould – *Penicillium notatum*.

The mould and its furry growth are commonly seen on old bread, greyish-green in colour and with a musty odour. For many centuries, different moulds had been used for medicinal purposes though with no understanding of how they worked. Ancient Egyptians and Greeks sometimes used mould to treat infections, and peasants of eastern Europe routinely applied mouldy bread or 'warm soil' to wounds as a traditional remedy. Even earlier, Indigenous Australians scraped mould off the sheltered side of eucalyptus trees for that purpose. Florey's first biographer, Lennard Bickel, relates how a bushman once brought a smelly bundle of moulds wrapped in sacking to the Walter and Eliza Hall Institute in Melbourne, suggesting they be investigated because they appeared to defeat infection and promote healing. Sir Charles Kellaway, the institute's director, told this to students in 1943, noting his regrets, in light of Florey's work, the moulds were thrown away, the man regarded as but a carrier of folk tales.

Among modern scientists, Louis Pasteur was among the first to recognise that moulds in the *Penicillium* genus inhibited the growth of bacteria, including animal bacteria harmful to humans such as anthrax. The discovery – or rather yet another rediscovery – of penicillin reads like a fairy tale. In 1928, Alexander Fleming, was, in his own words, 'play[ing] with microbes' in a small second-floor laboratory at St Mary's Hospital in London. As the story goes, one day he carelessly left the lid off a Petri dish and a mould spore landed in it contaminating the culture of bacteria. Returning from holidays, Fleming noticed that something 'funny' had happened to his staphylococcus culture: there was a zone around the mould which was completely devoid of bacteria. Elsewhere in the Petri dish, the bacteria flourished. Fleming concluded that the mould was producing a substance that killed the bacteria. The mould was later identified as *Penicillium notatum* and its active antibiotic substance was termed 'penicillin'.

Florey's American biographer has disputed the story, concluding 'it is impossible'. He persuasively argues that both the timing of Fleming's vacation and the sparse treatment of the incident in his notebooks condemn the well-worn story to myth. Moreover, the conventional fairy-tale account of Fleming's discovery of penicillin was not published until 1944, years after Florey and his team had proven the healing power of penicillin and founder's rights were in contest.

Whatever the truth of Fleming's account, this much can be said: Fleming's experience in fighting bacteria and treating infections led him to recognise the importance of his initial observation. He was the first to publicise the antibacterial qualities of penicillin, writing several papers in the late 1920s.

Fleming knew that penicillin was important. He knew it killed bacteria. He used his early mould broth as an antiseptic wash, and it worked to clear skin infection. He even knew it was not toxic when injected into animals. But having established the negative – that penicillin was *not harmful* to the body – there he left it, just one experiment short of one of the most significant medical breakthroughs of all time: to establish the positive – that penicillin was *effective* inside the body, to treat those infections beyond the grasp of surface antiseptics.

Fleming failed to pursue penicillin because he found it unstable and, as St Mary's did not have an experienced chemist on its staff, too difficult to grow, extract and purify in sufficient quantities. 'The trouble of making it seemed not worth while', he later recalled. In any case, he believed that it lost its potency when injected into the blood and tissue of living animals. It was interesting and promising, but it was in the end all too difficult – except, ironically, for cleaning bacteria off his Petri dishes, which he would continue to do.

By 1932, Fleming published his last work on penicillin and went on to other research. He had glimpsed the genesis of antibiotics but could not coax it from its mouldy broth, or 'mould juice', as he called it.

Splitting the atom: the bomb

On Wednesday, 27 April 1932, a new play by poet Robert Nichols and actor-producer Maurice Browne opened at London's Globe Theatre. Debuting just a few months before Stanley Baldwin's House of Commons lament about unstoppable bombers, *Wings over Europe* eerily foreshadowed much of Baldwin's air of fear and foreboding.

Set in a place Baldwin knew well, the Cabinet Room at No. 10 Downing Street, the play centres on a young scientific genius, Francis Lightfoot, who has learned how to unlock the energy in the atom. Lightfoot, who happens to be the prime minister's nephew, confronts the British Cabinet with the news that humankind is now capable of destroying civilisation or, if they have the wit and good will, creating a utopia. 'Today I put in your hands ... ultimate power over matter; the power of – of a god, to slay and to make alive', he announces to an appalled Cabinet.

Offstage, Lightfoot demonstrates the power of his discovery by having an assistant detonate a device the size of a lump of sugar which leaves a 'crater as big as St. Paul's'. Depressed by humanity and cynical politicians, Lightfoot threatens to destroy Britain unless the government agrees to international control of atomic power. In the nick of time, as the deadline nears, he is shot by the secretary of war.

But it's not over yet. News reaches the prime minister that the 'United Scientists of the World' have also learned the secrets of the atom. They send airplanes laden with atomic bombs over the world's capital cities threatening destruction unless an international congress is convened to ensure wise use of this new power. The curtain falls as the roar of airplanes fills the theatre – the bombers of the apocalypse are about to get through.

Only a few days after the premiere, on Sunday, 1 May 1932, readers of a popular London tabloid *Reynolds Illustrated News* opened their paper to a headline screaming 'Science's Greatest Discovery, The Atom Split at 100 000 volts, Secret of Cambridge Laboratory, Making a New World'. Newspapers reported that Cambridge scientists John Cockcroft and Ernest Walton working under Rutherford's direction had unleashed untold and unimaginable energy when they split the atom using a 'particle accelerator'. Reporters quickly surmised, as had the playwrights Nichols and Browne, that this energy might liberate humanity as easily as it might destroy it. 'Let it be split, so long as it does not explode', pleaded the *Daily Mirror* while the rival *Sunday Express* warily headlined, 'The Atom Split, But World Still Safe'. So prescient was *Wings over Europe* that there was popular confusion between the playwrights and the scientists. A play of science fiction at the beginning of the week had by the weekend evolved into an anxious portent of the future.

Unsurprisingly, newspapers captured the high drama but not the full story. Splitting the atom was in a sense old hat. Rutherford had first achieved this feat in Manchester back in 1917, albeit accidentally and without fully understanding what he had done, when he bombarded nitrogen atoms with radiation, causing them to shed particles and become oxygen atoms. Nitrogen atoms were in effect split, releasing energy in the process. Rutherford called it 'play[ing] marbles'.

The innocent schoolboy description masked the importance of his discovery: he had not only achieved the dreams of alchemists by transforming matter from one element to another, but also revealed the tantalising possibilities of energy contained within an atom.

《 • 》

Splitting an atom was one thing, being able to harness this energy from inside the atom, the way the power of steam and electricity had been harnessed over the previous two centuries, was another. It was potentially huge – on an atomic scale – but not so in practice, on a human one. Even at the artificially high speeds created by Cockcroft and Walton's particle accelerator, the positively charged protons still largely missed or were repelled by the positively charged nuclei. Scientists were expending a lot more energy firing protons than they got from occasionally splitting an atom. And so Rutherford, and his eminent contemporaries Niels Bohr and Albert Einstein, all concluded that exploiting atomic power on a meaningful scale was close to impossible. Rutherford called the prospect 'moonshine'.

In truth, most were happy that nuclear energy could not be easily extracted and controlled. Contemporary newspaper reports, while acknowledging the power locked in the atom, breathed a sigh of relief that humanity was still safe.

Theatregoers who nervously applauded *Wings over Europe* as a prescient work of imagination had their anxiety eased by comforting words from Cambridge's world-leading scientists: it was possible, but not practicable – science, yes, but also – as yet – fiction.

A scary novelty for the public, for scientists' concerns about the possibility of unleashing the power of the atom for destructive ends were far from new. As early as 1903, Rutherford mused that 'a wave of atomic disintegration might be started through matter, which would indeed make this old world vanish in smoke'. Years later, the horrors of the First World War left him hoping humanity would not discover the means of harnessing the atom's energy until it learned to live in peace.

He was not alone in his fear. In 1921, while seeking to explain the implications of Rutherford's research, German scientist and Nobel prize-winner Walter Nernst wrote, 'We may say that we are living on an island of guncotton'. He quickly added the (partly) consoling reflection, 'But, thank God, we have not found a match that will set light to it'.

Yet while great physicists doubted they ever could – and hoped they never would – one professional politician and amateur science enthusiast demurred. Writing a few months before the premiere of *Wings over Europe* and Cockcroft and Walton's breakthrough in splitting the atom, Winston Churchill was confident this 'match to set the bonfire alight, or it may be the detonator to cause the dynamite to explode' would soon be found. 'The discovery and control of such sources of power', he posited, 'would cause changes in human affairs incomparably greater than those produced by the steam engine four generations ago'.

The vision that Churchill painted of a nuclear-powered world was a near-utopian one, where almost nothing would be beyond humanity to better itself. Yet the realist in him could not but ponder:

> Explosive forces, energy, materials, machinery, will be available upon a scale which can annihilate whole nations. Despotisms and tyrannies will be able to prescribe the lives and even the wishes of their subjects in a manner never known since time began. If to these tremendous and awful powers is added the pitiless sub-human wickedness which we now see embodied in one of the most powerful reigning governments [that of Stalin], who shall say that the world itself will not be wrecked, or indeed that it ought not to be wrecked?

Churchill titled this prophetic 1931 essay 'Fifty Years Hence'. Little did he know that this dreaded future would arrive well ahead of his schedule, and he himself would have to deal with nuclear quandaries both times he served as prime minister.

《 • 》

As Churchill peered into his crystal ball – with another wicked regime, Hitler's, soon to materialise – 1932 was to deliver another crucial missing piece of the atomic puzzle. The problem was this: physicists could not explain why the same element could weigh different amounts, despite having the same constant number of protons and electrons. In 1920 Rutherford had speculated on the existence of yet another sub-atomic particle to explain this conundrum, but he could find no proof.

Another Cavendish colleague, James Chadwick, working from the early results of Cockcroft and Walton's particle accelerator experiments, finally found the elusive third particle. It was hiding inside the nucleus all along, alongside protons. He called it a neutron. The problem that had becalmed and frustrated nuclear physics finally had an explanation. There could be versions of an element with differing atomic masses, even though their structure and therefore chemical properties were the same. The number of extra neutrons explained the difference. Atoms of the same element but different weight – because they were packed with extra neutrons – were named isotopes.

Even more importantly, Chadwick established that neutrons were, as the name suggests, neutral; that is, they had no electric charge. And since they had no charge, they could hit an atom's nucleus without being repelled by the positive charge of the protons inside it. As the nucleus lay exposed to them, neutrons were therefore ideal to probe the secrets of different elements; they were the perfect missile to split the atom. In discovering the neutron, Chadwick had unwittingly found 'the match'.

But he didn't realise it. Chadwick and his immediate successors underestimated the power of the missile. While they knew a neutron could with relative ease hit the nucleus and cause it to shed some particles – an artificially induced decay or radiation – they did not believe it had enough energy to break the atomic nucleus completely apart.

Two years after Chadwick, Italian physicist Enrico Fermi began bombarding a range of different elements with neutrons. In virtually every case, the bombardment was causing atoms to shed particles, transmuting one element into another of differing weight. Fermi eventually bombarded uranium-238, at 92 protons and 143 neutrons the heaviest element found in nature, concluding two new, heavier elements with 93 and 94 protons each would be produced in the process.

He would soon realise he was wrong, though not before he was awarded a Nobel prize in Physics, in part for his supposed addition to the Periodic Table. What he had actually managed to do is split the uranium atom into two fragments of roughly equal mass, unwittingly achieving what a few years later would be described as 'nuclear fission'.

The work of Cockcroft, Walton, Chadwick and Fermi threw up another intriguing possibility: what if the missile strike was not a one-off? What if in splitting the atom, some of the released neutrons might in turn hit and split other atoms? If more than one neutron was released from each split atom, the process might expand exponentially. The result would be a chain reaction and the potential release of a vast, virtually unthinkable amount of energy. It was this revelation that finally made possible the concept of an awesomely powerful and destructive bomb.

The idea of a self-sustaining 'chain reaction' was first conceived by Leo Szilard, a Hungarian-born Jewish physicist with one of the most original and intuitive minds of the age. Daydreaming one autumn afternoon in 1933 at an intersection on Southampton Row in Bloomsbury, not far from the British Museum, Szilard stumbled upon an idea that would alter the frontiers of nuclear physics and change our world:

> As the light changed to green and I crossed the street, it ... suddenly occurred to me that if we could find an element which is split by neutrons and which could emit *two* neutrons when it absorbed *one* neutron, such an element, if assembled in sufficiently large mass, could sustain a nuclear chain reaction. I didn't see at the moment just how one would go about finding such an element or what experiments would be needed, but the idea never left me. In certain circumstances it might be possible to set up a nuclear chain reaction, liberate energy on an industrial scale, and construct atomic bombs.

Szilard was alarmed by the possibilities his mind had conjured. So certain was he that such a process might provide the scientific precondition for a new type of bomb, he patented it and in 1936 offered it free of charge to the British Admiralty for safekeeping. Not knowing quite what they had in their hands, the Admiralty accepted Szilard's gift and kept it secret in a vault until 1949.

But neither patent laws nor official secrecy could put the nuclear genie back in a bottle. Scientists in France, Italy, the United States as well as Britain were now all firing up particle accelerators in the hope of splitting atoms and releasing the energy inside.

《 • 》

In December 1938, less than a year before the outbreak of the Second World War, scientists in Nazi Germany achieved nuclear fission and, unlike Fermi, understood exactly what they had accomplished. Building on the Italian's work with uranium and borrowing Chadwick's technique for producing a particle beam, what Otto Hahn and Fritz Strassman observed during their experiment shocked them. Bombard most other elements with neutrons and their nuclei will change only slightly, losing or gaining a few protons; bombard uranium and its nuclei will dramatically split and break into two roughly equal pieces. But not only that. These two pieces will weigh much less than the original uranium nucleus – another insight that eluded Fermi. Where did the missing mass go? While the question was tantalising, Albert Einstein had already provided the answer.

It was $E=mc^2$, energy equals mass times the speed of light squared. Einstein's famous formula and revolutionary insight was that matter and energy were the same and interchangeable; that mass converted into corresponding energy and that energy converted into corresponding mass, and that you can easily calculate one if you know the other. The speed of light, the fastest speed we know in the universe, is 300 000 kilometres per second; squared it equals 90 000 000 000. Thus, the potential for enormous energy to be extracted from just tiny amounts of matter.

So in Hahn and Strassman's experiment, the mass lost when the uranium atom split in two was in fact converted into energy – enormous energy. It did not take long for this phenomenon to be explained. In Sweden, calculations by Austrian refugees from Nazism, Lise Meitner and her nephew Otto Frisch, showed that so much energy had been released by neutron bombardment of uranium nuclei that a previously undiscovered process must have been at work. Frisch named the process 'fission'. He even calculated that the energy released from splitting just one uranium atom could make a grain of sand make a visible jump. Imagine if trillions (and trillions) more atoms were fissioned.

Not only did each fission release enormous energy, but every split uranium nucleus released multiple neutrons. If these neutrons were to then smash into other nuclei releasing even more neutrons, in turn colliding with yet more nuclei, uranium might provide the answer to Szilard's daydream (or nightmare). Beginning with the splitting of just a single uranium nucleus and the release of energy and multiple neutrons, a 'chain reaction' might occur allowing rapid multiplication of one to ten to hundreds to thousands to millions to billions and then to trillions of trillions of energy releases – all this in a hot microsecond.

The realisation that with a sufficient mass of uranium atoms a chain reaction might become self-sustaining, overnight transformed the science of nuclear physics from the life of the mind and the practice of laboratories to the exercise of politics and the plaything of national security. Physics was no longer a purely academic adventure. For the possibility of a chain reaction brought with it not only the prospect of limitless energy but also the ultimate weapon hinted at by science fiction and predicted by *Wings over Europe*. The temptations of power – both military and electric – would now drive innocence from the soul of physics.

News of the work of the German and Austrian scientists spread quickly. A powerful new bomb was closer in theory than ever before, yet practical difficulties remained. The great Danish physicist, Niels Bohr, pointed out one of them: ordinary uranium (uranium-238) could not sustain a chain reaction. A very rare isotope known as uranium-235 could. No one yet knew how to separate it from the non-fissible bulk of uranium-238, but it was surely only a matter of time before scientists worked it out. The race was on.

Bizarrely, with war looming and despite physicists understanding the potential military application of nuclear fission and chain reaction, the customary scientific collegiality and openness remained. Throughout 1938 and 1939, as Hitler in quick succession snapped up Czechoslovakia's majority-German Sudetenland, the willing

Austria, and the rest of the unwilling but by then fatally weakened Czechoslovakia, Leo Szilard lobbied tirelessly and managed to convince leading physicists like Fermi as well as British scientists to self-censor so as not to assist any atomic-bomb program that the Nazis might have. But many wouldn't listen to Szilard's pleas. In France, Frederic Joliot-Curie, the son-in-law of radioactivity pioneers Marie and Pierre, refused to cooperate, enticing others to continue to publish sensitive research even as it became apparent that Hitler's aggressive instinct was far from sated.

In any case, despite Szilard's valiant efforts, it was already too late. In June 1939, just three months before general hostilities commenced, German scientist Siegfried Flugge published the results of a successful uranium chain reaction, concluding that 'our present knowledge makes it seem possible to build a "uranium device"'.

By now, the possibility of a uranium bomb was well acknowledged throughout the scientific world. In a far-sighted and prescient editorial for the journal *Discovery*, published in September 1939 just as Europe went to war, CP Snow wrote: 'Some physicists think that, within a few months, science will have produced for military use an explosive a million times more violent than dynamite. It is no secret; laboratories in the United States, Germany, France and England have been working on it feverishly since the Spring'. Snow explained that the 'principle' behind a 'uranium bomb' was 'fairly simple'. Though expressing his concern over whether men could be trusted with a new weapon of gigantic power, Snow did not doubt that the bomb should be built:

> Yet it must be made, if it really is a physical possibility. If it is not made in America this year, it may be next year in Germany. There is no ethical problem; if the invention is not prevented by physical laws, it will certainly be carried out somewhere in the world ... [W]e must not pretend. Such an invention will never be kept secret; the physical principles are too obvious, and within a year every big laboratory on earth would have come to the same result. For a short time, perhaps, the U.S. Government may have this power entrusted to it; but soon after it will be in less civilized hands.

As the Second World War commenced, the Germans seemed to have the upper hand. Hahn and Strassman's fission of the uranium atom and Flugge's account of a chain reaction gave the Third Reich the early lead in the race for a 'uranium device'. The foremost theoretical physicist of the day, Werner Heisenberg, was now leading the German work on nuclear fission. If Hitler was dangerous enough with his conventional armies, the Fuhrer in possession of the most powerful bomb ever devised was a prospect that did not bear contemplating.

For the British, it looked bleak. As Mark Oliphant later recalled: 'One was working against the clock. The British Intelligence evidence ... was that the Germans were well on the way'. As if exploring the possibility of microwave technology for radar was not enough for one person, a surprising turn of events would soon force Oliphant to take on another, much less official, mission. It would make him the crucial link in a chain of events leading from a provincial lab in the West Midlands, through the corridors of power on two continents, to the deserts of New Mexico, and finally the skies over Japan.

4

Prelude to war

Nestled among Oxford University's laboratories and libraries dedicated to scientific research is a handsome but unspectacular three-storey, red-brick building. Constructed in the Georgian style and looking like a metropolitan hospital, it was opened in 1927, thanks to the bequest of Sir William Dunn. The Victorian-era banker and merchant described in his lifetime as 'pathologically mean' but generous indeed in death.

This otherwise unremarkable building set on a leafy street near central Oxford is one of only three overseas places listed by the Australian Government as of outstanding historical significance to Australia. The others are Anzac Cove at Gallipoli and the Kokoda Track in Papua New Guinea. In contrast to the two bloody battlefields, the Oxford premises are dedicated 'to alleviate human suffering'. This is the Sir William Dunn School of Pathology and it is here you will find Howard Florey's laboratory.

Make your way through the large peaked front doors and climb the grand curved oak staircase, past the portraits of former heads of the school, and arrive at the top floor. Then, turn right along the corridor, walk about 20 metres, and you will arrive at the door to the graceful, but not grand, office of the director of the school. It sports a good view over the University Parks, with the River Cherwell beyond. In winter, the director might see a good game of rugby, and in summer perhaps a cricket match. This was Howard Florey's office. A short walk down the corridor are the teak laboratory benches where he worked. They have not changed much – except for a few coats of paint.

Florey took up his post as Professor of Pathology on 1 May 1935 and a little later was formally appointed Director of the Dunn School. In addition, the 'Statutory Professorship' came with a fellowship at the small and comfortable Lincoln College. But, despite Oxford's attractions, it was not an easy beginning. Staff morale was low and student numbers down. Remarkably, the Dunn School had an academic staff of just six, four of whom were part-time, assisted by five technicians. Student class numbers were ten or fifteen at most. On Florey's arrival at the Dunn School it was described as 'a mausoleum' with employees 'uneasily aware of their own slackness'. Given Florey's reputation for straight-talking, staff were understandably anxious about their futures.

But Florey moved with tact and caution. He was now in charge of one of the premier research posts in experimental pathology in the world. He had dreamt of this appointment for years and was intent on building the school up to its potential. For all the problems he inherited, it was still Oxford, and Oxford attracts stars.

Florey let some researchers remain, despite early reservations. He decided to give them a chance under his new leadership. The bacteriologist, AD Gardner, was 'much relieved' when asked to stay and went on to flourish under Florey, becoming a vital member of his team.

Having revamped the curriculum and attracted more funding and new students, Florey moved to recruit additional researchers of promise. This was, after all, his main focus. A large part of Florey's genius was his appreciation of scientific teamwork. A decade later, when accepting the Nobel prize, he would reflect that, 'Perhaps the most useful lesson which has come out of the work on penicillin has been the demonstration that success in this field depends on the development and co-ordinated use of technical methods'.

Florey's approach to recruitment was innovative and inspired. Because his research interests now focused on natural antibacterial substances, Florey was particularly keen to enlist a biochemist to investigate the properties of these substances. He soon found a brilliant one.

Ernst Boris Chain was just 29 when Florey, somewhat surprisingly, invited him to Oxford. A Jewish native of Berlin, he fled to Britain when the Nazis came to power, not so much out of fear, he later wrote, but disgust. A pianist of considerable talent with a possible career as a concert performer, Chain always looked part-flamboyant artist, part-scientific boffin. Reminding some of a young Einstein, this raffish and highly strung Continental would throughout his career confound his more sedate British colleagues.

His appointment to the Dunn School, following a stint at Cambridge, set in motion one of the most consequential relationships in modern medicine, one built upon entirely different temperaments and styles. Florey was reserved and spare, Chain voluble and romantic. Florey was tough but egalitarian, Chain emotional and hierarchical. Florey was a Puritan, Chain a Cavalier. Ebullient, nervous, generous and volatile, Chain proved a brilliant if sometimes brittle companion to Florey.

At first the two formed an unlikely bond, walking home through University Parks in the evening and sharing their problems and progress at work, and, occasionally, even private concerns. But, under the pressure of events and stress of wartime deadlines and secrecy, their friendship soured and, in time, they sustained no more than abrupt professional courtesies.

Chain commenced work, interestingly enough, on the biochemical effects of snake venom and then the metabolism of cancer tissue. This latter work required the assistance of another biochemist with ability in micro-techniques – measuring the sometimes minute quantities of chemicals in organic substances. Chain recalled Norman Heatley's work in Cambridge and recommended him to Florey. Clutching his newly minted Cambridge doctorate, 25-year-old Heatley arrived in Oxford in September 1936. Dark haired, thin and taciturn, Heatley soon proved indispensable to the team's work, where his technical ingenuity shone. Florey charged Dr Margaret Jennings, another important addition, with work in pathology. Florey's team grew over the next few years. With the advent of war, Ethel Florey began to work full-time after the departure of their children to the safety of the United States.

Inevitably, perhaps, Chain's relationship with Norman Heatley soured too. Chain, as Heatley's nominal superior in the lab, took him for granted and treated him with condescension. The courtly and less experienced Heatley could not so easily fight back. Such niceties did not worry Florey who would always give as good as he got. Chain recalled 'the very walls of Florey's office would shudder with our shouting'. Yet this odd couple – truthfully, the whole odd team – would show that personal compatibility is not a prerequisite for success and that a diversity of temperaments, traits and tactics can make for a momentous breakthrough.

Impressed by Chain's work, Florey asked him to investigate Alexander Fleming's lysozyme, a naturally occurring antibacterial substance found, among other places, in human tears. Chain would collar Florey on their walks home and give excited updates on his work. When it proved a scientific dead end, they both pondered one day, why stop at lysozyme? Why not take a look at other naturally occurring antibacterial substances – especially the antibacterial substances known to be produced by microorganisms, such as yeasts or fungi? We might just find something useful there. Chain jumped at the chance and started his research in the manner of every good scientist – with a review of the literature on the topic to help narrow down the best prospects for further investigation.

Chain was thorough. His review listed about two hundred references on substances with antibacterial properties. What struck Chain was how little was known about them. He was convinced of the necessity of further research. In early 1938, leafing through the library stacks of the Dunn School, he picked up a volume of the *British Journal of Experimental Pathology*. Chain had consulted this journal before, noting relevant articles by, among others, Fleming and Florey. But this time he picked up volume 10, published in 1929, and, flicking through the bound journal, spied on page 226 an article titled 'On the Antibacterial Action of Cultures of a Penicillium With Special Reference to Their Use in the Isolation of B. Influenzae'.

Chain's dark eyes squinted as he read: 'A good sample will completely inhibit staphylococci, *Streptococcus pyogenes* and pneumococcus in a dilution of 1 in 800 ... The toxicity to animals ... appears to be very low ... Constant irrigation of large infected surfaces in man was not accompanied by any toxic symptoms ...'

Chain's interest was immediate; the strong effect against staphylococcus in particular caught his attention.

Then 'something seemed to click' in Chain's mind: he had recently come across the mould described by Fleming.

But where? Was it by any chance in a Petri dish that he saw his colleague, Margaret Campbell-Renton, carrying along the corridor that separated their laboratories? Not waiting a moment, he sought her out. A surprised Campbell-Renton replied that, yes, it was. Compounding the serendipity, she revealed to Chain that Alexander Fleming himself had given a sample of the mould to Florey's predecessor, Professor Georges Dreyer. Hardly believing his luck, Chain raced the Petri dish back to his lab and began the first tentative experiments to investigate the mysterious mould's healing qualities.

Florey now had to take a gamble. Lysozyme had not been the success they had hoped for. Penicillin mould held some promise, but there were other contenders in the mix, including *Bacillus pyocyaneus* and *Subtilis-mesentericus*. Every new avenue of research carried its own considerable risks. A decision had to be made.

Paquita, Florey's daughter, later recalled a walk through the park near the Dunn School with her father when they came to a chestnut tree. 'This is the tree', said Florey, 'under which we used to talk very often, Chain and myself, and under this tree we came to the decision that of the possibilities for pursuing research we would choose Penicillium, and so the decision was arrived at'.

But the decision was not easy. Florey made his fears known to a fellow Australian at Oxford, Dr Roy Douglas Wright (much better known as 'Pansy'), telling him, 'There is no question we will now have to go for penicillin. But let's face it; it is obvious that penicillin is a tough project. Some very fine chemists have tried to isolate it and have got nowhere. It could be just another blind alley, and where would I be then'.

Isolating the active substance in the mould had indeed defeated all others, from Fleming onwards. But Chain was confident. He believed that others had failed because they weren't skilled chemists. What swayed Florey was his familiarity with the substance. He had been aware of penicillin for years and its possibilities had been scratching away at him as he kept himself acquainted with sketchy research by Fleming's successors. Florey had even been one of the editors of the 1929 *British Journal of Experimental Pathology* that included Fleming's penicillin article that so captivated Chain, though could not recall reading it at the time.

Still, it was chancy. Fleming and his few successors had got nowhere. If Florey followed them to a dead-end it would be difficult to hold his strong team together. As war loomed, the practical demands made on universities and laboratories would be irresistible, forcing current research to give way to military needs. On a personal level, another failure like lysozyme would detract from his own professional prestige and affect his standing with donors.

A recurring theme in the story of penicillin, and scientific research more broadly, is the never-ending quest for money. The Dunn School received about half of its funding from the university, another quarter from the Dunn Endowment Fund, and the balance from student fees and other grants. It was never enough. Florey's role as fundraiser was the bane of his existence, and a major distraction from the scientific work he loved, yet it was absolutely essential to further his team's work. No money, no science. But he resented the task. In the summer of 1939, he wrote complaining to Sir Edward Mellanby, Secretary of the Medical Research Council and a firm supporter of Florey's, that he was 'fed up' with the 'difficulties of trying to keep work going here'. Tempted yet again to invoke the epithet of the Aussie bushranger, he worried that he had 'acquired a reputation of being some sort of academic highway robber – I have to make such frequent approaches for grants'.

While Florey loathed lobbying, he did become cannier at writing grant applications, to bodies both in Britain and the United States. He began emphasising practical medical or therapeutic outcomes rather than simply 'expanding scientific knowledge' – even though, between themselves, Florey and Chain were motivated more by scientific interest than any deep humanitarian instincts. Florey's appreciation of the politics of grant-giving was to prove a winning attribute.

《 • 》

Oliphant arrived at Birmingham in the shadow of Rutherford's death. He had lost a father figure and a mentor; the sheer joy of work at the Cavendish was gone now too. It was quite an adjustment for both Oliphant and Rosa as they arrived in the industrial Midlands city of Birmingham. It might have been Britain's second largest metropolis, but it lacked the intellectual and cultural gravity of other big cities. Called 'the first manufacturing town in the world', Birmingham did at least have a history of scientific and technological innovation, from James Watt's steam engine onward. As became its industrial character, the city was grey and gritty and the university campus in no way comparable with Cambridge. The Physics Department, which Oliphant was to head, was moribund, not unlike the Dunn School upon Florey's arrival.

But the news was not all bad. As Oliphant got down to work, the Vice Chancellor was supportive and money for research soon came Oliphant's way. Lord Nuffield, founder of Morris Motors, a car company with a large factory located in Birmingham, donated £60 000 pounds (equivalent to about £4.3 million today). Frustrated by Rutherford's penny-pinching at the Cavendish, Oliphant was keen to embark on innovative research at his new university. Not least, he wanted a cyclotron – a particle accelerator – that would allow much easier penetration of the atomic nucleus. Oliphant was certain it would foster great new discoveries in atomic physics.

But before investing a fortune in building his own cyclotron, Oliphant needed to investigate other efforts in this area of experimental research. So, with the Vice Chancellor's approval in late 1938, Oliphant travelled to San Francisco to examine the cyclotron at the Berkeley campus of the University of California, under the direction of Ernest Lawrence. A year before, Lawrence became an unwitting pioneer of nuclear medicine when, in a flash of inspiration, he and his physician brother, John, directed the cyclotron's beam at their dying mother and managed to kill her abdominal cancer. Gunda Lawrence went on to live another 22 years, the first successful case of cancer radiation therapy in history. Intrigued by its possibilities, Oliphant would later consult Florey about the use of the cyclotron for medical purposes.

Oliphant had briefly met Lawrence before, on the latter's European trip in 1933, but now, with more time, they basked in each other's company. Oliphant's fact-finding expedition to America cemented a friendship that was to have a direct bearing on the United States' commitment to developing an atomic bomb.

On his return to Birmingham in early 1939, and revelling in his experience at Berkeley, Oliphant announced he would build a 60-inch cyclotron, the same size as Lawrence's. It would be housed in a pit underneath a purpose-built, small single-story structure. Public speculation, even alarm, grew as materials arrived and work progressed.

To build a respected new physics department, Oliphant needed the best researchers, collaborators and staff. He was given great authority by the university to hire and exercised that authority wisely, drawing together experienced academic physicists and others just completing their doctorates. History would prove his choices inspired, and a formerly moribund department was soon at the forefront of the wartime technological race.

Back in the spring of 1937, while still at the Cavendish, Oliphant pulled aside Rudolf Peierls, a Jewish physicist and refugee from Hitler's Germany. They had met twice over the years and Peierls was taken with the Australian's zest, inventiveness and informality. Oliphant, in turn, enjoyed Peierls' company and respected his intellect. Dark haired, diffident and slight, Peierls was gentle and unassuming, yet by the time of his death in 1995, he would be described as 'a major player in the drama of the irruption of nuclear physics into world affairs'. Now Oliphant asked him: 'What would you say if I tried to organise for you a chair of theoretical physics in Birmingham?'

Despite strong competition from, among others, fellow Australian Harrie Massey, Peierls got the job. He was forever grateful to Oliphant for putting his name forward. After four years in limbo, unable and unwilling to return to Hitler's Germany but not yet a British citizen (he became one in 1940), Peierls was now a professor, with the security and independence of a permanent university position (the newly created department was called Applied Mathematics). He soon repaid Oliphant and Britain for their kindness and refuge.

A trip to Denmark also proved valuable in the recruitment stakes. In early 1939, Oliphant visited the University of Copenhagen's Institute for Theoretical Physics, presided over by Niels Bohr. Bohr had just returned from a conference in Washington DC where, to a stunned international audience, he broke the news that Otto Hahn and Fritz Strassmann, scientists working in Nazi Germany, had split the uranium atom.

'Fission' and the enormous release of energy it produced heralded the arrival of the atomic age. With Europe on the brink of another war, the timing could not have been more ominous. But the nuclear scientist in Oliphant was fascinated by the news. He first read the report in the 2 February issue of *Nature* and quickly made his own back-of-an-envelope calculation: if you split every atom in a pound of uranium (a big 'if', to be sure) you might produce an explosion equivalent to about 50 000 tons of conventional explosives – as if about 100 000 average-size General Purpose aerial bombs went off at once. It was staggering – and intriguing. For the first time Oliphant contemplated the terrible possibility of an atomic bomb. But atom bomb or, preferably, no bomb, he hoped to move to the forefront of this new frontier of physics research.

He did not have to go far. Right there, at Bohr's Institute, was 34-year-old Otto Frisch, who just a few weeks before, in partnership with his renowned physicist aunt, Lise Meitner, had confirmed the enormous energy latent in the uranium atom. They provided the theoretical proof of 'nuclear fission' – a term that Frisch himself coined.

Oliphant already knew the earnest Frisch by reputation and was in no doubt that he might conduct valuable research and attract prestige to his department's fission program, particularly once the cyclotron was operational. Like Peierls, Frisch was a brilliant young Jewish scientist, whose promising future was denied to him under Germany's anti-Semitic laws. He was now looking for a safe new home to live and work.

Oliphant was keen to secure Frisch's services, though he had no positions available. So he offered Frisch a job that would justify his relocation to Britain for at least a few months: 'You just come over in the summer. We'll find you something to do. You can give a few lectures or something. We can discuss the matter'. After that, who knew? The world might be at war. They would work something out.

For Frisch, this was a lifeline. He quickly took up Oliphant's written offer of employment confirming his oral assurance in Copenhagen. Despite a lack of job security Frisch knew he was better off in Britain. 'I had a fear', Frisch later recalled, 'that Denmark would soon be overrun by Hitler, and if so', hoped for 'a chance for me to go to England in time, because I'd rather work for England than do nothing or be compelled in some way or other to work for Hitler or be sent to a concentration camp or gas chamber'.

Quickly, but without fuss, Frisch journeyed to England carrying with him just two small suitcases and dreams of resuming his life's work. He arrived in the English summer and the university break and spent much of his time in the sun with Rudolf Peierls, his new Birmingham colleague, and Peierls' wife, Genia. 'There was very little else to do', he later recalled. It would be his last quiet summer for a long while.

« • »

Oliphant's new lab was coming together. He was building the largest cyclotron in Britain and attracted ambitious young scientists to his cause. His international reputation was growing, no longer eclipsed by Ernest Rutherford's brilliant star. Oliphant was excited by recent discoveries in nuclear physics, including nuclear fission, and confident his new department would soon play a leading role in atomic research. With inspired recruitment, assertive administration and shameless fundraising, his future and that of his department looked assured.

But then, in Oliphant's words, 'all was overtaken by the war'. With a call from Whitehall, Oliphant's laboratory, like many others, was drafted for war work, to assist in the defence of the British realm. Pondering the mysteries of the atomic nucleus merely for the wonder, glory and beauty of science were over.

It was his old friend from the Cavendish Laboratory, John Cockcroft, who inducted Oliphant into the top-secret world of military research. Oliphant was by now, as the English put it, part of the 'old boy network'. He had been of vital assistance to Cockcroft and Walton in building their particle accelerator that had so spectacularly 'split the atom' in 1932, and for which the duo would later be awarded the Nobel prize. Oliphant would later describe Cockcroft as 'a man of action rather than of ideas' who wrote 'few scientific papers'. Somewhat unkind, perhaps, but not inaccurate. Tall and well-built and a fine sportsman when young, Cockcroft had a distinguished career as a science administrator, including as Chancellor of the Australian National University between 1961 and 1965 (to be followed in that role by Howard Florey).

Sir Henry Tizard had courted Cockcroft, as a First World War veteran and eminent researcher, to convince other scientists, in particular physicists, to assist in the war effort. Tizard was the scientific advisor to the Air Ministry and the government's point man for science at war. Over lunch at London's Athenaeum Club in early 1938, Tizard revealed to Cockcroft the jealously guarded secret of long-wave radar and the development of the Chain Home stations, Britain's radar shield facing the Continent. Cockcroft recalled Tizard telling him, 'These devices would be troublesome and would require a team of nurses. Would we – the Cavendish Laboratory – undertake to come in and act as nursemaids, if and when war broke out. He talked also of wanting large powers at short wave lengths and would we think about it'.

And so, when Cockcroft approached Oliphant, it was not nuclear research he wanted to discuss with his old colleague. In the autumn of 1938, Oliphant was ushered into Tizard's secluded office in the Air Ministry where Tizard and Cockcroft briefed him about the progress in radar research and the daunting problems that remained. Oliphant was stunned by these revelations but excited at the prospect of being included in this top-secret research.

Oliphant readily understood the problem. The 10-metre-long radio waves could cover much of the sky – a wide, 100-degree slice in front of the station – but had little ability to discriminate targets. Chain radars were a 'floodlight', but Britain now desperately needed a more discerning 'searchlight' to better spot the enemy in both defence and attack. The shorter the wavelength, the more information might be gathered about a target. Microwaves – measured in centimetres rather than metres – were known to possess greater range and directional accuracy. They would be able to eliminate the distortions of 'ground clutter', caused by long waves bouncing off the ground. They would also be able to locate much smaller objects and distinguish between closely bunched targets. Most importantly, microwaves would require smaller devices. An onboard radar, in turn, would give pilots 'night vision'.

Cockcroft and Tizard now briefed Oliphant that Edward 'Taffy' Bowen, a young physicist on Watson-Watt's team, had been working on ways to make radar airborne. 'I did a back-of-an-envelope calculation of the best wavelength for the purpose', Bowen recalled later. To eliminate 'ground clutter', a narrow beam the width of 10 degrees would be required. 'Given an aperture of 30 inches – the maximum available in the nose of a fighter – this called for an operating wavelength of 10 centimetres'.

By 1939 Bowen and his team had succeeded in reducing the wavelength from 10 metres to around 1.5 metres. It was a huge feat and miraculous in such a short time. But the necessary antenna (now in the shape of dish-like mirror to better direct radio waves in a narrow beam), at 1.5 metres in length, while small enough for larger aircraft and naval vessels, was still too big to fit on a fighter. For that to happen, it would have to shrink by another 90 per cent. With the Luftwaffe gearing for war, there was little time left.

So how to best use the time? Opinions were divided. Some, especially the RAF, wanted scientists to pursue microwaves, come what may. But, as Bowen later recalled, 'there was practically no interest within Bawdsey Manor itself and anyone talking about centimetre waves was thought of as some kind of crank'. For most scientists, microwaves were a pie in the sky. Worse, they were a distraction. For Watson-Watt and others, the top priority was to ensure that what they already had – Chain Home – worked as best as it could to stop German bombers. Any time left, and all available hands, should be committed to this.

And so, initially at least, the scientists would concentrate on being 'nursemaids' for the existing Chain Home technology rather than midwives of further invention. Over the first few months of 1939, Cockcroft, under Tizard's instructions, recruited more than 80 physicists to spend a month working on-site at radar stations up and down Britain's coast. Six universities, including Oxford, Cambridge and London, sent teams to assist. Oliphant's Birmingham team was assigned Britain's southernmost radar installation, Ventnor on the Isle of Wight. Located about a kilometre out of town on St Boniface Down, with steep slopes down to the English Channel, the RAF Ventnor station was ideally placed to detect incoming aircraft from Europe. For that reason, both the town and the radar station were bombed within months of Oliphant's visit.

Yet as Cockcroft's teams dispersed to their designated stations to iron out technical problems before fighting started, it proved impossible to keep inquisitive minds off the challenge of designing the next generation of smaller and more versatile radar. It would not be easy; hardly any of the few dozen scientific recruits – and none of the Birmingham men – had a background in radar technology. For Oliphant it was a plus: he figured they would not be limited by existing know-how or inhibited by prior convictions about what might or might not be possible. Oliphant later reflected that he put together a group of people who would 'think in fundamental terms rather than just follow radio practice'.

Upon arriving on the Isle of Wight in late August 1939, Oliphant commandeered an empty hotel. He briefed his team on existing radar technology and challenged them to improve its performance. He led the way, poring over blueprints and tinkering with parts, trying to coax more out of the giant sentinel peering into the skies over Normandy. They did all they could in a short time, much shorter as it turned out than they had hoped. Ventnor was ready for action, as ready as it could be within the limits of existing technology.

Only a few days into their inspection, on 1 September, Nazi Germany invaded Poland. Two days later Britain and France declared war on Germany. Oliphant beat a hasty retreat to Birmingham where he assigned research tasks. But two of his staff, a young research fellow, John Randall, and a graduate student, Harry Boot, stayed behind on the Isle of Wight to finish the team's work, in the process gaining deeper insights into the challenges ahead.

Having been able to closely observe long radio waves at work, Randall and Boot understood intimately their power but also their limitations. 'By now', they recalled many years later, 'we were all interested in the possibility of producing microwaves'. The prospect of being able to do so 'for airborne and shipborne use' – the holy grail that Tizard was after – 'was most exciting, but apparently impossible. So we all returned to Birmingham to try'. The possible was already done, the impossible would start tomorrow.

5

Hell breaks loose

On 3 September, at 11:15 am, Oliphant and his team crowded around their hotel radio on the Isle of Wight to listen to the grim words of the prime minister, Neville Chamberlain.

Just 12 months before, Chamberlain had arrived back from Munich a hero, having secured Hitler's promise that, in exchange for Czechoslovakia's Sudetenland, Germany would make no further territorial demands. Gleefully waving the agreement signed by the Fuhrer, Chamberlain declared, 'It is peace for our time'. *The Times* swooned, 'No conqueror returning from the battlefield has come home adorned with nobler laurels'.

Now, speaking from the Cabinet room at No. 10 Downing Street to a nation shocked but resigned to fight, Chamberlain announced that Britain was again at war with Germany. Hitler had lied. Nazi Germany first swallowed the rest of Czechoslovakia back in March, and then, on 1 September, invaded Poland, Britain's ally. 'You can imagine what a bitter blow it is to me that all my long struggle to win peace has failed', Chamberlain added. It was a blow from which he would never recover.

As the national anthem sounded following the prime minister's pained announcement, Oliphant exchanged uneasy glances with his men. The Ventnor radar station could become essential any day now. Would the Chain Home system stand the test of combat? And would Britain's best physicists have enough time to develop the next generation of radar needed to give Britain a clear edge over the Luftwaffe squadrons? Or would the bomber get through? Nothing like national survival to add urgency to one's work.

« • »

Three days earlier, at 8 pm, on 31 August 1939, Kurt Hartmann, Luftwaffe Lieutenant with the 77th squadron of Stuka dive bombers stationed at Neudorf airfield just across the border from Poland, hastily scribbled in his diary, '*4.15 Uhr geht es los. Gott sei dank, endlich eine Entscheidung*' – At 4:15 am we begin. Thank God, decision at last.

Just after 4:30 am, and without any warning, Hartmann and his comrades dropped 46 tons of explosives and incendiaries on the small town of Wielun in central Poland. Ninety per cent of the city centre was destroyed in the raid, including a clearly marked hospital, a church and a synagogue, killing 127 residents, many in their sleep. Wielun had no military targets. Defenceless civilians were the first victims of the Second World War, setting the stage for the next six bloody years.

Hitler had ordered more than half of the 4000 Luftwaffe fleet to support his invading armies. Facing them, the Polish air force could muster only 600 airplanes, mostly old models. Despite the odds, Polish pilots managed to inflict heavy losses. But individual bravery could only go so far; from the second week of September the Luftwaffe enjoyed virtually unchallenged air superiority.

The grisly consequences were plain for all to see, in part thanks to vivid dispatches by reporters on the ground and often under fire. Relentlessly bombed from the air and reduced to rubble by artillery, Warsaw surrendered on 28 September. If Guernica was a prelude, Poland was the main show; the first large-scale demonstration of modern air power, feared by so many for so long, from HG Wells to Stanley Baldwin. They were right. The bomber got through.

After five relentless weeks of apocalypse from the air, a Nazi radio broadcaster taunted his foreign listeners, 'I should like the gentlemen of London to see what a city looks like when it has been through what Warsaw suffered. These gentlemen ought to see what might happen in their own country if they persist in their mad warmongering'. Shattered Poland surrendered three days later. The gentlemen of London were indeed paying attention and what they saw were their own worst nightmares.

«•»

Haunted still by the mechanised slaughter in the trenches a quarter of a century before, neither friend nor foe greeted the coming of a new war with much enthusiasm.

The Fuhrer, himself among the millions caught up in the patriotic fervour of August 1914, was disappointed that Berliners did not take to the streets to cheer the announcement of war with Poland. Two days later, the American journalist William Shirer reported that the people of Berlin heard the news that Britain and France had declared war against Germany in 'shocked silence'. Hitler had overestimated Germany's enthusiasm and underestimated Britain's resolve. But he did not relent.

On the boulevards of Paris, the mood was likewise subdued, if not gloomy. In London it was resolute but sombre and quiet. A few minutes after Neville Chamberlain's address to the nation, air raid sirens sounded throughout Greater London and much of the country's south-east. It was a false alarm, but within days British life changed in anticipation of imminent aerial bombing. Military planners had estimated 20 000 casualties in the first 24 hours and up to 150 000 in the first week of a German air offensive. Keith Hancock recalled a civil servant's warning, 'The British public must not be led to expect that there will be coffins for everybody'. Sandbags were stacked around public buildings, night-time blackouts began, and above the major cities familiar fleets of barrage balloons once again floated in the skies. 'When the sun is low and strikes below them early in the morning and at sunset they gleam like quicksilver, their fins shadowy – fish in a deep-blue sea', wrote one Londoner to her American friend. To underline the seriousness of the new normal, all entertainment venues were closed indefinitely. Fifty engineers from the infant television service were redeployed to work on radar projects after it too was closed down.

But the expected German air offensive failed to materialise. Hitler would fight Britain and France on his own timetable: defeat Poland first and secure the Reich's new eastern border with the Soviet Union; and then turn west with the full force of his arms. As the Luftwaffe bombed Polish towns and villages, strafing columns of soldiers and refugees throughout September, the RAF responded by dropping thousands of leaflets over Hamburg, Bremen and the Ruhr, warning German civilians about the danger of an all-out war. No casualties were reported on the ground from the falling bundles of paper.

The public reaction across Britain was mixed: shame at the failure to meaningfully assist their Polish ally, guilty relief that the war had not yet touched home, resignation that it would, and, once that happened, grim determination to finish the business with Germany once and for all. But for now, an uneasy calm descended upon Britain; as one historian pithily put it, 'All the sensations of war, in fact, except the fighting'. The Phoney War had begun.

《 • 》

Universities too shared the rigours of preparing for war.

Unlike Oxford's dreaming spires, Birmingham was a strategic target for German bombers. With its large Spitfire factory and the famed Birmingham Small Arms plant, the city was the industrial heartland of Britain. The so-called Birmingham Blitz commenced on 9 August 1940, the last bombs dropping on 23 April 1943. Only London and Liverpool were bombed more heavily. When war was declared, the university's Edgbaston campus readied itself by blacking out windows, abandoning night classes and giving over student accommodation for war purposes. Everyone from undergraduates to professors signed up to serve in the Home Guard or Birmingham City Fire Service.

Oliphant hated war but his pacifist instincts were overwhelmed by a certainty that Nazi Germany represented a singular and unambiguous evil. He wanted to employ science to defeat Hitler. He would do his part.

Having already devoted some of the previous 12 months to the problem of radar, in September 1939 Oliphant secured a contract from Charles Wright, the director of scientific research for the Admiralty. Oxford was also contracted, as was Cambridge, which replaced the University of Bristol in the competition. Wright was an early Cavendish alumnus, a member of Robert Scott's South Pole expedition in 1911, and subsequently a talented science administrator. He had been impressed with Oliphant's decisive modernisation of the Birmingham laboratory and judged him capable of quickly adapting to radar work. The Birmingham outsiders would be now challenging Oxbridge's scientific domination. 'And not only outsiders, but colonials!' Wright's demand to all three universities was straightforward yet daunting: 'give us many watts on few centimetres'. In other words, make microwave radar a reality.

Oliphant hated putting his beloved cyclotron project on hold but with Britain in peril, personal research interests would take a back seat. He moved quickly to give tasks to independently minded scientists. Some grizzled at their new roles. All were assigned to groups of two or three, sometimes to improve current technology, others to imagine new ways. John Randall and Harry Boot arrived back in Birmingham full of ideas and enthusiasm sparked by their time on the Isle of Wight, but were disappointed to learn that Oliphant had given them a research assignment with apparently less likelihood of success.

At the outbreak of war, there were a few known ways to generate microwaves but none suitable for battle-ready radar. The most promising was the 'klystron' invented in 1937 by American brothers Russell and Sigurd Varian at Stanford University. While Oliphant had seen an early version of this device during his visit to the United States in 1938, the invention was only publicised in May 1939. The klystron was essentially a vacuum tube with a cathode at one end, from which an electron beam was shot to amplify a high-frequency radio signal passing through the tube.

This immediately led to excited speculation about its application to radar. The klystron certainly generated microwaves, but the electron beam could not draw sufficient energy to guarantee the strong and consistent output needed for long-distance radar detection. Still, the invention was encouraging, and Oliphant devoted most of his resources to enhancing this device. Soon, his improved klystron could produce 400 watts of power. Its 7-centimetre waves successfully detected an aluminium-painted, gas-filled barrage balloon, which at one point during the test lifted Oliphant off the ground and carried him 100 metres down a field. As exhilarating as Oliphant's flight was, detecting a balloon tethered to the ground – or to a scientist – was one thing, detecting German airplanes long before they reached British airspace quite another, and even at 400 watts, the klystron's output was not nearly enough for the task.

Another method for generating microwaves was the magnetron. The first simple prototype was built in 1920 in the United States and subsequently replicated independently over the next two decades in several other countries. This device also consisted of a glass vacuum tube. The tube was surrounded by a magnetic field, which caused the electrons shot from a cathode to follow curved, as opposed to straight, paths along the tube. The magnetron could produce microwaves of a centimetre in length, at 30 or 40 watts, just about enough to power a weak light bulb. Only a tenth as strong as Oliphant's improved klystron, it seemed unlikely ever to power workable radar.

This was the device passed to the unimpressed Randall and Boot. John Randall was 34, with a creative mind and discipline for hard work. He was originally inspired by the Adelaide-born Nobel laureate Lawrence Bragg to train as a physicist (the world of science still being relatively small, their paths would cross again after the war, when Randall's team at King's College collaborated with James Watson and Francis Crick from the Cavendish Laboratory, under Bragg's leadership, to successfully model the structure of DNA). Harry Boot, 12 years younger than Randall, was still working towards a doctorate under Oliphant's supervision. Young and impetuous, he was particularly annoyed at being assigned an unpromising project.

Marooned in a small room behind a lecture theatre, Randall and Boot built their own magnetron and pondered how to increase the meagre power output. More than 35 years later the two physicists recalled: 'Fortunately we did not have the time to survey all the published papers on magnetrons or we would have become completely confused by the multiplicity of theories of operation'. Unburdened by the furrows of previous practice and theoretical clutter, Randall and Boot asked themselves instead if the best features of both the klystron and the magnetron might be combined to produce a superior next-generation device.

In thinking through the problem, Randall recalled rummaging through a second-hand bookstore on his recent summer holidays with his family in the Welsh seaside town of Aberystwyth. There he came across a copy of HM Macdonald's seminal 1902 book *Electrical Waves*. Not typical summer holiday reading perhaps, but it sparked Randall's imagination.

In the book, Macdonald detailed the great German physicist Heinrich Hertz and his famous electric 'spark-gap experiment'. Hertz was the first to produce high-frequency radio waves and then, as they oscillated through a loop of copper wire, detect them in the form of a spark jumping a minute gap in the loop. While Hertz's observation inspired the first radio transmitters, the wire technology only generated short pulses of radio waves, which restricted it largely to sending Morse Code. The technology was superseded in the early 1920s by vacuum tube transmitters able to produce the continuous waves used to carry sound in contemporary radios. But Randall was nevertheless intrigued. If high-frequency radio waves could be generated across a gap in a wire, might it be possible to generate them in three dimensions, in other words, across the gap inside a metal cylinder?

Randall's musty book also detailed a calculation that caught his eye. Macdonald showed that the length of radio waves produced would be 7.94 times the diameter of Hertz's loop. So, asked Randall, would that also apply to the diameter of a metal cylinder?

He and Boot decided to find out. To achieve Oliphant's target of a transmitter that would produce 10-centimetre wavelengths, thought to be optimal for radar, they rounded up Macdonald's figure to 8 and calculated that the diameter of the tube would have to be about 1.2 centimetres.

The next step was pure genius. One afternoon in November 1939, the young men had an idea: why only one

1.2-centimetre-wide tube – why not a number of them, say six, placed around an electron beam–producing cathode? The power output would be multiplied, generating much stronger waves. Randall started to scratch a design on the back of an envelope while Boot did the calculations to create the first hazy sketch of arguably the most consequential invention of the Second World War. Watson-Watt would later call the document 'The Magnetron Memorandum'. It was 'a communal manifesto for the Centimetric Revolution in Radar'.

With its six cylindrical 'cavities' drilled through a copper block equidistant around a central cathode, Randall's rough design of the cavity magnetron looked more like a Colt revolver than a radio-wave transmitter. In fact, Colt revolver chambers were initially used as a drilling jig and, later on in the United States, machines used to cut Colt cylinders were repurposed to make the magnetron's copper core element. But there was nothing of the Wild West about the complex physics involved.

Oliphant was still pinning his hopes on the klystron and trying to modify it to produce sufficient power for workable microwave radar. In his enthusiasm, he had co-opted most of the staff, space and resources. Grabbing a lift home with Oliphant one evening, Randall outlined the progress he and Boot had unexpectedly made with the magnetron. He cautiously sounded out Oliphant about some additional bench space and money. Quickly recognising the potential of his younger colleagues' revolutionary idea, Oliphant found the room and the cash, and became increasingly interested in their work. He gave them a free hand as they started the difficult task of building their cavity magnetron prototype.

Through Christmas 1939, Randall and Boot laboured day and night over their secret design. They paused only once, to move to a newly built facility that Oliphant had helped secure, the Nuffield Research Laboratory. As quickly as they could, they gathered the necessary equipment. Oliphant arranged for a couple of transformers and rectifiers to convert the common alternating current into the direct current needed to power the new machine.

Boot's workshop skills were now put to the test. He did not let Randall down. Oliphant's interest might have been piqued, but he was by no means convinced about the likelihood of success. While he provided all the resources he could spare for them to build the prototype, Randall and Boot had to beg, borrow and steal the rest. Slowly, their creation came into being. When a metal disk was needed to plug one end of the primitive apparatus, Boot used a halfpenny. Joints were crudely sealed with wax. A Birmingham scrap-metal yard soon became their most important supplier. It was a time of scarcity, but they found a way. With its echoes of 'string and sealing wax' improvisation, Ernest Rutherford would surely have approved. The war had made an old man's penny-pinching prejudice everyone's necessity.

By February 1940, after months of intense work, Randall and Boot were ready to test the first model of their cavity magnetron. It resembled a small engine rotor with a series of cylindrical holes cut into it and was crafted from a solid block of copper attached to an untidy tangle of tubes and pumps. A car headlamp was connected to demonstrate the output of power. On the morning of 21 February, with the flick of a switch, the cavity magnetron burst into life and the headlamp shone spectacularly. While the original simple magnetrons could barely power a light bulb, the new resonant cavity magnetron was soon producing so much power it burned out headlamps and lit cigarettes. Oliphant, who joined the excited group within half an hour of the switch being flicked, quickly ordered the headlamp replaced with hardier fluorescent tubes. They too lit up the room – as well as the spirits of Randall, Boot and Oliphant.

On 24 February, Boot confirmed with a scrawl in his black laboratory notebook that the radio wavelength produced by the resonant cavity magnetron was 9.5 centimetres – almost exactly what the Royal Air Force needed for the next generation of portable, all-seeing microwave radar. Power output was soon confirmed at an extraordinary 400 watts – ten times better than any previous magnetron. And while 400 watts appeared to be the ceiling for the klystron, for the cavity magnetron it was just the beginning.

Randall and Boot, late arrivals from the Isle of Wight charged with a secondary line of research and experiment, had created the central component of a powerful new radar device that would give the British and the Allies a decisive advantage in the Second World War. Other people had previously developed a cavity magnetron, as the device came to be known, but Randall and Boot were the first to build a model powerful enough to be of practical use.

The two young scientists always credited Oliphant's 'great drive' for the success of the entire unlikely enterprise. It was, they reflected years later, 'fortunate ... that we were in Oliphant's laboratory at the University of Birmingham'. Oliphant was project manager, mentor, fundraiser and problem solver. Not a natural diplomat, he nevertheless reached out to those who might help and provided his team with the necessary resources. Always available, always questioning, if not always right, Oliphant inspired and lifted his staff to achieve ends that just six months before were pie in the sky. He was an internationally renowned scientist in his own right, but, just as importantly, a talented science administrator, with a keen eye for fresh talent and the ability – as well as the dogged persistence – to make things happen even under the most challenging conditions. Without Oliphant's leadership, drive and intuition in those crucial months of 1939 and 1940, the development of modern microwave radar would have been delayed, with dismal consequences for the defence of Britain, its airspace and its sea lifelines to the world.

The February breakthrough was promising, but the next test was more daunting: to turn a laboratory marvel into a weapon of war – to take an experimental prototype, a mass of tubes, pumps, wires and copper, and refine and refashion it into a fully operational system capable of mass production and application on the frontline.

In March the Admiralty experts left a demonstration underwhelmed by technical bugs plaguing the early model and a few weeks later brought in General Electric & Co at Wembley. The practical technical expertise they brought to the project soon proved invaluable. In little time their engineers succeeded in improving the overall design, made the device airtight, did away with bulky pumps, and replaced water with air cooling, making the magnetron more practical for use onboard aircraft. There was still a way to go before microwave radar could debut in combat, and even longer before there were sufficient devices to make an impact. In the meantime, the first-generation long-wave radar of Chain Home dotted along England's east coast remained the only means of radio detection. Britain was expecting bombers at any moment: in a few weeks, Home Chain would be put to the test.

《 • 》

As the war began, feats of improvisation and ingenuity playing out in Oliphant's lab were being replicated 90 kilometres south in Oxford, where another scientist in his 20s was being forced to use old bathtubs, bedpans, food tins and milk churns to grow penicillin mould. 'At one period the ordinary empty soft drink bottle became a prized possession', Norman Heatley later recalled. At this early stage of the penicillin project, Heatley's job was first and foremost. Sir Henry Harris, an Australian who succeeded Florey as head of the Dunn School, put it this way: 'Without Fleming, no

Chain or Florey; without Chain, no Florey; without Florey, no Heatley; without Heatley, no penicillin'. If Heatley could not grow enough mould, Chain would have nothing to extract and purify, and Florey nothing to test.

It nearly didn't happen. Heatley, tired of his mistreatment by the temperamental Chain and preparing for a possible career in industry, was awarded a Rockefeller Foundation travelling grant to study biochemistry and hone his skills in micro-manipulation at the Carlsberg Laboratory in Copenhagen, not far from Niels Bohr's Institute of Theoretical Physics, from which Otto Frisch had recently fled. Heatley was planning to depart on 11 September but when war broke out, he decided to remain in Oxford. It was a rare stroke of luck; Florey knew that without Heatley's skills the whole project would certainly be delayed, if not defeated.

Like at Birmingham, scientists at the Dunn School also prepared for the bombing of their university and the likely invasion of the British Isles. A photograph of the time shows Heatley in shirtsleeves digging an air-raid shelter in the school's backyard. Scientists, students and support staff all pitched in. And Florey, a keen photographer, recorded much of it as sandbags were stacked outside the school's walls and windows were darkened to ensure blackouts. Everywhere they went, staff carried gas masks, concerned that the Germans might unleash chemical weapons, as they had in the previous war. Florey himself agreed to give up a third of his time, if necessary, to 'poison-gas research', while Chain headed a blood-transfusion service headquartered at the

Dunn School and linked to local hospitals. Like Oliphant's, Florey's team would also play their part.

Fortunately, Oxford escaped bombing and the damage and destruction visited on so many other cities during the war. One theory speculates that Hitler planned to use Oxford as his capital after he conquered Britain. Another postulates a 'gentlemen's deal' of some sort where Germany eschewed bombing Oxford or Cambridge if the British spared Heidelberg and other old German university towns. Sadly, there is no proof of such a civilised deal. Oxford probably survived unscathed since Luftwaffe pilots would have to fly over London and its formidable defences to reach it.

When Britain declared war on Germany, Florey was finalising his first major funding proposal to pursue penicillin research. He had thrown his lot in with the mysterious mould and went all out to impress the Medical Research Council and its Secretary (and former colleague at the University of Sheffield) Sir Edward Mellanby. Uncharacteristically optimistic – Florey's future colleague and biographer Gwyn Macfarlane later speculated that for once 'Chain's enthusiasm overcame Florey's scientific caution' – his proposal claimed that 'Penicillin can easily be prepared in large amounts ... In our opinion the purification of penicillin can be carried out easily and rapidly'. This would have been news to Norman Heatley. Florey also stressed the potentially great therapeutic value of the project, proposing penicillin be injected intravenously in order to study its 'antiseptic action' directly *'in vivo'*, not just in a Petri dish.

Florey's uncharacteristic rhapsodising worked, after a fashion. Pending a full examination of his proposal by the Council, Mellanby forwarded £25 'for expenses'. Under a separate grant, Chain was also awarded £300 with £100 for expenses per annum for three years thus, if nothing else, providing him with security of employment.

So, with £25 and a dash of hope, a revolution in human healing began. It was a start, but it was nowhere near enough. And Florey, like so many research scientists before and since, grew sick of the endless rounds of funding applications and increasingly self-conscious of his emerging status as an 'academic highway robber'. But the powerful team he assembled could do little without the necessary equipment and materials.

Not for the last time in the penicillin story, Florey would also reach out for help to America. As a graduate at Cambridge, he had benefited from a Rockefeller Foundation travelling grant to the United States, and not long after taking up his post at Oxford, he received a Rockefeller Foundation grant for equipment. But now his request was for much more: enough to underwrite his team's research for the duration of the project.

Salvation came in the person of Harry M 'Dusty' Miller, one of the Rockefeller Foundation's representatives in Europe. Miller was a fluent French speaker based in Paris, oozing American optimism and a can-do attitude. Like other 'circuit riders', his task was to travel the scientific world in search of projects worthy of support by the foundation, underwritten by the fortune of the Standard Oil founder, John D Rockefeller Sr.

Miller's diaries from his numerous circuit rides illustrate the importance to science of these talent spotters with their seemingly bottomless moneybags. In June 1939, for instance, Miller was in Copenhagen, offering assistance to Niels Bohr's Institute of Theoretical Physics and to Otto Frisch's aunt, Lise Meitner, marooned in Stockholm but attempting to retrieve her books and furniture from Germany (which she subsequently did).

On 1 November 1939, the globe-trotting Miller was in Oxford, having lunch with none other than the university's Vice Chancellor, Professor George Gordon, and his family. For an American with money, no doors were closed. Dusty Miller, an early scout for an ascending America, embodied the changing fortunes of Anglo-American relations.

After lunch, Miller dropped in to see Florey at the Dunn School. He found the lab 'quite full' with evacuees from the Ministry of Health in London, but other than that, his timing could not have been better.

The straight-talking Florey mentioned to Miller, a trained bacteriologist, that he was 'just on the point' of asking the Rockefeller Foundation for a grant to support research he was beginning on 'naturally produced bacteriological inhibitors'. Miller's ears pricked with interest. He respected Florey, noting in a memorandum sent a few days later to the Rockefeller Foundation headquarters in New York that the young Australian is 'practically the only experimental pathologist of any real distinction in the British Isles'. Miller added that Florey 'has the full confidence of Mellanby'. Assured of Florey's skills and intrigued by the penicillin project, Miller recommended that the request for support ('maximum would certainly not exceed 1500 pounds a year for three years', in Florey's calculations) 'should be given serious study'.

Miller's initial enthusiasm paved the way for Florey and Chain's formal application to the foundation a couple of weeks later, on 20 November. If approved, it would single-handedly take care of all the money woes and save Florey from the distraction of constant fundraising. With so much riding on its success, Florey ended his application letter with a new, if characteristically understated, appreciation for practical outcomes, something much favoured by the business-like Americans: 'It may also be pointed out that the work proposed, in addition to its theoretical importance, may have practical value for therapeutic purposes'.

On 19 February – two days before Randall and Boot successfully demonstrated their cavity magnetron to Oliphant in Birmingham – Florey and Chain learned that their application had been successful. They received £1250 or US$5000, guaranteed for a year but eventually extended to five. For both men, this changed everything, and they jumped for joy at this 'royal generosity'. The pressure of constant applications was gone; Florey and Chain could now focus solely on science. It was a game changer; Chain would later see it as the decisive moment in the sequence of events leading to the development of a miracle new drug.

The work began in earnest. Florey's claim to Mellanby that penicillin could be easily produced in large amounts would now haunt Heatley. Heatley developed a test to measure the potency of each batch. He tried many different recipes for the broth to grow penicillin mould, aiming always to produce more of it and of a higher potency. Sugars, salts, malts, meats and alcohol were all thrown in. Heatley even tried brewer's yeast and Marmite, the ghastly tar-like black sandwich spread much beloved by the English at breakfast. It was a messy, smelly, thankless task.

And frustrating. The problem of finding the optimal growth medium was one thing but a more prosaic lack of laboratory space and suitable vessels also restricted production and the quantities remained small. On top of this, extracting penicillin from the mould was extraordinarily difficult. Initially, ether was the only solvent known to filter out penicillin from the mould, but the penicillin disappeared when it was removed from the ether. This vanishing trick had confounded and defeated Fleming and his successors. Chain, fascinated by penicillin's instability, threw himself at the problem. He developed techniques of stabilising penicillin at certain pH (acidity) levels and extracting it at just above freezing. Even so, it took hundreds of litres of mould filtrate to produce only minute quantities of penicillin itself.

The first critical breakthrough in logistics came on 18 March 1940. Heatley called his new method 'reverse extraction'. Building on previous practice, it used ether to extract penicillin from the mould (at Chain's suggestion amyl acetate soon replaced ether and proved a better and safer solvent) and then dissolved it in water. At last, Chain now had enough material to play with and purify. But he never forgave Heatley for succeeding where he had failed.

Assisted by Edward Abraham, another young biochemist who joined the team in November 1939, production, extraction and purification proceeded in leaps and bounds. Abraham contributed a newly discovered technique of alumina column chromatography, which greatly assisted in removing impurities, and Chain, using another recently discovered technology – freeze-drying – was able store the penicillin without sacrificing its potency. After weeks of work, Chain held a test tube with a few pinches of a dusty brown powder – penicillin's first face to the world.

Chain was keen to start testing penicillin on animals to determine its toxicity. He had already diluted a milligram of the brown powder in a litre of distilled water to discover it inhibited the growth of staphylococci bacteria. With such high potency, Chain would not have been surprised if penicillin proved toxic to laboratory mice. Busy with other responsibilities across the Dunn School, and with their relationship increasingly strained, Florey fobbed off Chain's immediate urgings. Chain proceeded undeterred.

The following day, without obtaining Florey's permission, Chain asked his colleague, Dr John Barnes, to inject about 40 milligrams of the brown penicillin powder mixed with a small quantity of water into the abdominal linings of two mice. Chain waited for any signs of distress from the rodents. After several hours, to his great delight, the mice showed no ill effects. Penicillin was not toxic.

Chain would later claim this experiment was 'the crucial day in the whole development of penicillin and the day on which everything became possible to us'. Not quite. Fleming had already shown in 1929 that penicillin was not dangerous to a mouse and a rabbit. Meanwhile, miffed that Chain went ahead with the experiment in his absence, Florey repeated it himself by injecting the tail vein of a mouse. By chance, the mouse urinated soon after. It was brown, the colour of the powder. With not a moment to miss, Florey took a drop of the rodent's urine and deposited it on a bacterial culture in a Petri dish. The urine soon killed the germs, proving that penicillin's antibacterial effect was not affected as it passed thought the mouse's body. The next question would go to the crux of their research project: what if the mouse injected with penicillin happened to be suffering from a bacterial infection? The mouse would not be harmed, this much seemed certain, but would the bacteria?

A revolution in healing, slowly inching ahead over the course of millennia, was now breaking into a sprint.

《 • 》

Not everyone at Oliphant's Birmingham laboratory was chasing microwave radar. Some were not allowed to.

For Frisch and Peierls, the languid, carefree days of summer were over. With Hitler triumphant over Poland

and Europe now at war, their worst fears were coming true. At least in Britain they remained out of Nazi reach.

They were also at a loose end. As 'enemy aliens' – Peierls became a British citizen in February 1940 and Frisch not until 1943 – they were prohibited from working on radar or any other top-secret project of importance to the war effort. In fact, they were lucky not be interned together with other German and Italian nationals. Never mind that they were refugees from Nazism and its anti-Semitic policies, British security regulations were a blunt instrument that did not discriminate between sympathisers and opponents of the fascist regimes.

Oliphant decided to leave the two talented young minds doing what they did best: investigating the possibilities of nuclear fission. This research was not considered of military significance and therefore not a state secret. Oliphant gave Frisch and Peierls an old lecture room to set up their desks and research equipment as they scrounged around for other resources. They commenced work only a few metres away from where Randall and Boot were simultaneously building their first cavity magnetron. Frisch and Peierls always remembered Oliphant had brought them together and set their course. He was, Frisch later said, 'absolutely a visionary'.

In September, Oliphant wrote to Tizard seeking more uranium oxide for experiments. Tizard was not impressed: 'I am told that you have refugees in your Laboratory', he wrote, 'and it just occurs to me that perhaps it is a little hard on the English physicists who are interested in the same problem but who are now deeply engaged on war work'. Oliphant replied that the research would be undertaken by Dr Otto Frisch who, with his aunt Lise Meitner, had only a year earlier provided the theoretical explanation of nuclear fission. He then added forcefully: 'In my opinion it is much more important that work of this nature should be done than that any question be raised about whose effort is employed to get the answer'. Tizard relented, and Birmingham got the uranium oxide.

Oliphant was notoriously lax in security matters. Over tea he would sidle up to Peierls and with a sheepish grin ask him a 'hypothetical' question, which Peierls would dutifully take time to consider and return later with an answer. As Frisch recalled, 'Peierls knew that this was connected with the generation of very short electric waves, such as were needed for radar, and Oliphant knew that Peierls knew, and I think Peierls knew that Oliphant knew that he knew. But neither of them let on'. For Frisch and Peierls, however, microwaves were a sideshow, splitting atoms the main game.

In a paper published on the day Nazi troops invaded Poland, Niels Bohr and John Wheeler identified the main problem facing nuclear scientists: being able to separate the scarce but fissionable isotope, uranium-235 (constituting 0.7 per cent of uranium), from the plentiful and non-fissionable uranium-238 (99.3 per cent). Wheeler later recalled that it was 'so preposterous then to think of separating U235 that I cannot forget the words that Bohr used in speaking about it: "It would take the entire efforts of a country to make a bomb'". At Imperial College, physicist George Thomson (son of JJ Thomson, discoverer of the electron) and his team were of the same opinion, and by February 1940, they too had virtually given up.

While the great Niels Bohr could not be easily dismissed, the worrisome possibility of a Nazi bomb kept nagging at Peierls. Earlier, in that last summer of peace, he tried calculating the 'critical mass' of uranium necessary to sustain a chain reaction and, thus, a nuclear explosion. His calculations showed that the quantity of natural uranium required would be 'of the order of tons'. Such a bomb would be too heavy to fit on a plane so an explosive atomic device, while not impossible, remained impractical. Still, Peierls had doubts about publishing his calculations in a paper. It would certainly fall into German hands and eventually could have a bearing on the design of a weapon. He consulted Frisch, then freshly arrived from Denmark, who 'saw no reason against having my paper published, since Bohr had shown that an atomic bomb was not a realistic proposition'. Fittingly, perhaps, rather than a physics journal, Peierls published his paper in the more obscure *Proceedings* of the Cambridge Philosophical Society.

A few months later, Oliphant tasked Frisch with reviewing recent progress in nuclear science for the British Chemical Society. Oliphant, disturbed by the possibility of a German bomb like his two young colleagues, was also grappling with the straw in the wind of nuclear weapons. In a pre-emptive strike at Bohr and Wheeler's soon-to-be-published article, Oliphant spelled out his concern in a letter to Bohr in late May 1939 that 'there may be no possibility missed that here is a possible source of power, or of explosion':

> I understand ... that you do not anticipate that the fission process will prove an explosive one, even under the best possible conditions, but it is felt here that the whole possibility must be investigated experimentally. As there are possibilities that the isotopes of uranium may be separated in reasonable quantity in the near future, perhaps this investigation is not so futile as may seem at first.

Half a year later, dressed in his winter coat in his freezing bedsitter, Frisch's numb fingers were typing the requested summary of contemporary research in nuclear science. He agreed with Bohr, concluding that 'the construction of a super bomb would be, if not impossible, then at least prohibitively expensive and that furthermore the bomb would not be so effective as was thought at first'. In fact, it would be 'no worse than setting fire to a similar quantity of old-fashioned gunpowder', Frisch later wrote recalling his thinking at the time. A nuclear chain reaction might well be possible but would not be fast enough to set off a super bomb, since the commonplace uranium-238 isotope would suck up most of the neutrons released, leaving only the very small amount of the uranium-235 isotope undergoing fission. It would be a fizzer. Even if it might be possible in the future to separate the isotopes, vast quantities of uranium would be required to extract a meaningful measure of uranium-235. It was not certain that such industrial quantities of uranium were even available, and besides, how much of the fissionable uranium-235 would you even need to create a self-sustaining chain reaction? No one knew that either.

But by the time his paper was published in early 1940 by the British Chemical Society, Frisch was having uneasy second thoughts.

For one, the recent theoretical advances in 'thermal diffusion', discovered by the German scientist Klaus Clusius, suggested separating the lighter fissionable uranium-235 from the heavier uranium-238 was finally possible. Frisch now asked himself the obvious question: precisely how much uranium-235 would be necessary to create an explosive chain reaction? Employing a formula previously used and refined by Peierls, Frisch made quick calculations. Expecting it would be a 'matter of tons', to his amazement the figures showed it was 'something like a pound or two'. Surely, he thought, this must be a mistake.

Frisch sought out Peierls, startling him with the blunt question, 'Suppose someone gave you a quantity of pure 235 isotope of uranium – what would happen?' Peierls, unassuming and unflappable, knew what his friend was after: how much uranium-235 would be needed at a minimum to start a self-sustaining chain reaction, the so-called 'critical mass'? They checked Frisch's earlier calculations. He was right: not the many tons, as all previous estimates claimed, but 'about a pound'. This turned out to be too low, but as

Peierls later pointed out, the 'order of magnitude was right'. In any case, what mattered was this: it was impractical to carry a bomb weighing tens of tons in an airplane; one weighing in pounds might be carried in a backpack.

But just how powerful would such a bomb be? Conversing excitedly in their native German and working – literally – on the back of an envelope, the answer came as a revelation: the 'critical mass' of a pound or so of uranium-235 would spark a chain reaction releasing 'the equivalent of thousands of tons of ordinary explosive', or TNT. 'We were quite staggered by these results', recalled Peierls, 'an atomic bomb was possible, after all, at least in principle!' They dubbed it 'the jitterbug'.

In the winter-chilled English Midlands, Frisch and Peierls laid bare a deadly secret of nature. But they were not so arrogant as to believe they could be the only ones. 'Any competent nuclear physicist', Peierls reflected later, 'would have come out with very similar answers to ours if he had been asked … The only unusual thing that Frisch and I did at this point was to ask these questions'. This was not a happy realisation. 'We were at war, and the idea was reasonably obvious; very probably some German scientists had had the same idea and were working on it', Frisch reasoned. Fission and fascism were a frightening combination. Fear would now drive nuclear research. Speed and secrecy were vital.

'Look', a chastened Frisch said to his friend as they contemplated the scribbled envelope in front of them, 'shouldn't somebody know about that?' Not knowing 'how to send a secret communication, or, for that matter, where to send it', there was only one person they trusted sufficiently, who would understand the physics and would know what to do.

Preoccupied as he was with the quest to develop microwave radar at his lab, Oliphant had little time to ponder nuclear science, his true scientific love and the primary research interests of his prewar career. Even so, he remained engaged, seeking uranium oxide for Frisch and Peierls to conduct fission experiments, reading with interest their publications and encouraging them to write more. As his letter the previous summer to Niels Bohr makes clear, Oliphant was not completely convinced by the scientific consensus that an atomic bomb was not feasible. So, in February 1940, when Frisch and Peierls nervously mentioned they now had serious doubts about the conclusions reached in their recent papers, Oliphant queried them. Intrigued and a touch anxious, he asked that they type up their findings and come back as soon as possible.

These two gentle, humane men, refugees from Nazism, left Oliphant's office to draft one of the most far-reaching scientific documents in history. Peierls typed up their conclusions on his old typewriter on foolscap paper and, sensing the importance and profound security implications of their work, made just one carbon copy. Like Randall and Boot's magnetron memorandum, the simple form of the document belied its revolutionary consequences.

Their paper, written in English, not their native German, is a model of clarity and simplicity. One half, 'On the Construction of a "Super-bomb"; based on a Nuclear Chain Reaction in Uranium', is a technical document. It is a blueprint for an atomic weapon. Such a weapon was feasible, Frisch and Peierls now thought, because of 'a possibility which seems to have been overlooked in ... earlier discussions', namely that the amount of fissionable material (uranium-235) needed for an atomic bomb – the critical mass – was much less than previously thought. Reworking their original calculations, they concluded that a 5-kilogram bomb would 'be equivalent to that of several thousand tons of dynamite'. While separating the uranium-235 isotope from uranium-238 would be expensive, producing the fissionable material could be achieved by existing techniques. Again, while Frisch and Peierls proved overly optimistic, in principle they were right. In any case, while the technical challenges were real, the weapon would be devastating and protection from its blast and radioactive fallout 'hardly possible'. The potential made it worth the effort. In that they were right too.

The second part of the document was titled 'Memorandum on the Properties of a Radioactive "Super-bomb"'. In many ways it is even more momentous, providing a glimpse into the minds of two scientists grappling with the consequences of their revelations. When Prometheus stole fire from the gods and gave it to humanity, Zeus punished him by chaining him to a mountain and having an eagle tear out his liver anew every day. They might get to keep their livers, but Frisch and Peierls foresaw the moral torments that nuclear scientists would suffer if this new power were unleashed. Conventional bombs allowed at least the pretence of targeting military objectives. The destructiveness of atomic weapons obliterated any distinction between a soldier and a non-combatant. As Frisch and Peierls wrote, 'Owing to the spreading of radioactive substances with the wind, the bomb could probably not be used without killing large numbers of civilians, and this may make it unsuitable as a weapon for use by this country'. Such qualms would soon become another casualty of total war. As Peierls reflected shortly before his death in 1995:

> In say 1937 everybody was horrified by the idea of air raids on cities killing civilians. And then a few years later you have the air raids on Hamburg, Dresden and Tokyo and now no one turned a hair. It was simply that the public attitudes had changed, and without that change it might not have been so easy to justify the dropping of the atom bomb.

After the Blitz, few cared any more.

In any case, it was potentially a matter of national survival; the prospect of Hitler getting his hands on an atom bomb before Britain did not bear contemplating. But it did not stop Frisch from worrying. In a letter to his mother, he wrote that he felt like a man who 'had caught an elephant in the jungle by the tail and who did not know what to do with it. But he knew it was an elephant'.

Fortunately, Oliphant knew what to do with the elephant. Escaping for a moment from the heady excitement of his team's success with the cavity magnetron two weeks earlier, Oliphant withdrew to the quiet of his comfortable wood-lined office to read the document that Frisch and Peierls had prepared for him. He was gripped and disturbed at what he read. He understood the memorandum's assumptions and calculations, grasping its momentous significance. Re-reading it several times, he rechecked the calculations, half-hoping his two colleagues might be in obvious error. They were not.

Oliphant now sat uneasily pondering where science meets war. Right now, he alone understood both the physics before him and the politics that lay ahead: how to get the 'powers that be' to stop and listen and act on this unexpected and potentially earth-shattering development. Ever practical – Lorna Arnold, an official historian of the UK Atomic Energy Commission, would later describe him as 'street-wise' – Oliphant understood that what might be obvious to scientists might not be obvious to laypeople. Even seemingly brilliant and self-evident ideas, concepts and inventions are not automatically guaranteed a hearing, much less success, in the absence of energetic and persistent advocacy – and not a bit of luck.

Oliphant was the right man in the right place at the right time. His standing could not have been higher in bureaucratic circles. Under his direction, radar would soon be radically improved to give the Royal Air Force a vital offensive edge in the air. Coupled with an outgoing (some would later say, rather too outgoing) personality and a forceful presence, Oliphant was someone to be reckoned with. He commanded attention.

First up, the most important individual for Oliphant was Sir Henry Tizard, chairman of the Committee for the Scientific Survey of Air Defence and godfather of the radar project. Tizard was a sceptic, reckoning the possibility of a nuclear weapon in the short term – the current war – at 100 000 to 1. Well aware of the difficulties but keen that the memorandum be given urgent and proper attention, Oliphant wrote a covering letter:

> I have considered these suggestions in some detail and have had considerable discussion with the authors, with the result that I am convinced that the whole thing must be taken rather seriously, if only to make sure that the other side are not occupied in the production of such a bomb at the present time.

On 19 March 1940, the Frisch–Peierls memorandum, with Oliphant's covering letter, landed on Tizard's desk. Tizard knew Oliphant well and appreciated his significant contribution to the war effort. Conclusions drawn in Oliphant's lab by physicist colleagues, and endorsed by the man himself, were sure to command Tizard's attention.

They did. If Frisch and Peierls were correct in their assessment, the political and military implications would be immense. Having read the memorandum, Tizard concluded there was little time to waste.

'What I should like', he replied to Oliphant, 'would be to have quite a small committee to sit soon to advise what ought to be done, who should do it, and where it should be done, and I suggest that you, [George] Thomson, and say [Patrick] Blackett, would form a sufficient nucleus for such a committee'. Persuaded of the importance and urgency of the task ahead, Tizard was pulling together some of Britain's leading physicists to the project.

Oliphant's first job, however, was sensitive and personal. He had been taken into Frisch and Peierls' confidence but could not reciprocate that trust. The atomic bomb was now a military secret. Oliphant told Frisch and Peierls that while their memorandum was being considered, he could not indulge their curiosity further. Their status as 'enemy aliens' or 'ex-enemy aliens' (in Peierls' case) had seen to that. Unable to work on radar, the duo had now written themselves out of their nuclear physics work. Before long, good sense prevailed and Frisch and Peierls would provide important technical advice in developing the bomb. But for the moment, they had to wait. From their pokey office in Birmingham, they had deftly calculated the building of an airborne atomic bomb. Their safe and unclassified theoretical work was no longer theoretical and no longer theirs.

The new committee met informally for the first time on 10 April 1940. A day earlier, Hitler invaded Denmark and Norway. The Phoney War in the West was over.

Young Howard Florey in the 1920s –
'striking good looks as well as a sharp mind'.

Young Mark Oliphant in the 1920s – 'gentle giant and benign boffin'.

Sir Ernest Rutherford's Cavendish Laboratory, ca. 1926
– secrets of the atom revealed with 'string and sealing wax'.

Niels Bohr Institute Conference, Copenhagen, 1937. Picture includes: Niels Bohr (front row, first from the left), Werner Heisenberg (front row, second from the left), Lise Meitner (front row, third from the right), Rudolf Peierls (second row, fourth from the left), Otto Frisch (fifth row, fourth from the right) and Mark Oliphant (back row, hands under chin).

Florey during the war – from Petri dish to miracle drug.

Oliphant in the United States, 1941 – about to dive into America's bureaucratic swamp to convince the US to build the bomb.

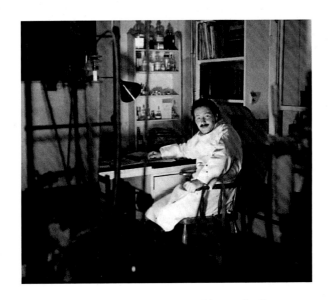

Florey's penicillin team: Ernst Chain – 'brilliant if sometimes brittle' chemist.

Florey's penicillin team: Norman Heatley – 'thin and taciturn' technical genius.

Oliphant's atomic bomb team: Otto Frisch (left) and Rudolf Peierls – prohibited from working on radar, their memorandum proved the atomic bomb feasible.

Oliphant's cavity magnetron team: Harry Boot (left) and John Randall celebrating their microwave prototype 26 years later – 'with its six cylindrical cavities it looked more like a Colt revolver than a radio wave transmitter'.

Friends in high places: Alfred Newton Richards once described Florey as a 'rough colonial genius'. One official history gave Richards 'the greatest credit in making penicillin available for use during the war'.

Oliphant, John Cockcroft and Robert Oppenheimer - 'the father of the bomb'. Oliphant first brought Oppenheimer into the atomic bomb project; he never escaped its fame or its burden.

America's scientific establishment in 1940, jovial and complacent, enjoying their last season of peace and quiet before Oliphant's arrival. From left: Ernest Lawrence, Arthur Compton, Vannevar Bush, James Conant, Karl Compton, Alfred Loomis.

Oliphant and Lawrence at Berkeley in September 1941, in front of the cyclotron but talking about the bomb – 'They were as alike as two peas in a pod'.

Dunn School's 'Penicillin Girls' – 'young, many only teenagers, poorly paid but enthusiastic and proud of their vital role'. Without them, Florey and his team could never have produced enough penicillin for human trials.

Mary Hunt (left) at work at the National Regional Research Laboratory in Peoria in 1943 – the mouldy melon she discovered was 'perhaps the most significant piece of fruit in history', its new strain ultimately yielding 1000 times more penicillin.

Dr Ethel Florey, Howard's wife – she was lauded for leading the second human trials of penicillin. 'It was hectic but she loved it', and achieved results that exceeded everyone's best hope.

Advertisement for penicillin production from *Life* magazine, 14 August 1944. By D-day there was enough penicillin to treat all military casualties. Civilians had to wait another year.

Fruits of Birmingham: the first airborne radar of the war, late 1941 – an AI Mark VIIIA unit in the nose of a Bristol Beaufighter Mark VIF night fighter. A German commander complained that the RAF was picking his Luftwaffe planes off 'like currants in a cake'.

Fruits of Birmingham: the first atomic bomb test at the Trinity site, New Mexico, 16 July 1945 – 'No amount of imagination could have given us a taste of the real thing'.

Sir Alexander Fleming, 'elfin and inscrutable', his discovery of penicillin mould in the 1920s inspired Florey and his team to develop penicillin the drug more than a decade later.

An edited picture worth more than a 1000 words: The Nobel Prize for Medicine in 1945 was jointly awarded to (from left) Florey, Chain and Fleming for 'the discovery of penicillin and its curative effect in various infectious diseases'. To media editors with a marking pen there was but one true winner.

Australian National University founders: Oliphant as Director of the Research School of Physical Sciences, 1950–1963.

Australian National University founders: Florey as Chancellor, 1965–1968.

Australian National University founders: (from left) Oliphant, Keith Hancock and Florey on the site of the future ANU campus in 1948.

Maestros on the mountain: Imagining a national university: (from left) Keith Hancock, Oliphant and Florey.

6

Sunday, 26 May 1940

From early May a political storm had been brewing in Westminster. After months of the Phoney War, Britain and Germany both decided to occupy Norway. Their aim was the same: to block the other's access to Sweden's mineral wealth, so crucial for armament industries. The Germans got there first, by a nose. The British, once they landed in mid-April, were outgunned, outmanoeuvred and isolated. It was the first direct clash with German arms since 1918 and it was a fiasco.

Some called the abortive invasion 'Churchill's second Gallipoli'. He was the principal architect and advocate of the Norwegian operation and, as First Lord of the Admiralty, oversaw the seaborne invasion. Its failings should have sunk Churchill's career as surely as the failure of the Dardanelles landings in 1915 halted his early rise in politics. 'Mr Churchill's sun has been called to set very rapidly by the situation in Norway', American Ambassador Joe Kennedy cabled President Roosevelt, expressing the accepted wisdom of the day. That the public somehow ended up blaming Prime Minister Chamberlain, and the Scandinavian debacle provided Churchill with a springboard to the highest office, is perhaps one of the more providential – if somewhat ironic – turns of world history.

On 7 and 8 May, parliament sat to debate the already doomed Norwegian campaign, but it rapidly became a complete review of the government's record in the war so far. Chamberlain, sick with undiagnosed terminal cancer, emerged as weak, indecisive and vulnerable. A prime minister so diminished might briefly survive in a time of peace and plenty, if backed by a united party. But Britain was at war and Chamberlain was attacked not just by the opposition, as might be expected, but by many of his own Conservative colleagues. Having lost the confidence of both parliament and the public, he had to go.

In this moment of crisis, a wartime National Government that included the opposition Labour Party offered the only way forward for the country. With Chamberlain no longer acceptable, who would lead it? Chamberlain called a meeting for the afternoon of 9 May at No. 10 Downing Street, inviting the contenders: Churchill and the Foreign Secretary, Lord Halifax. Chamberlain, the Conservative Party leadership and even the king favoured Halifax. But the Tory backbench and Labour MPs would not have 'the Holy Fox', an appeaser like his mentor Chamberlain, and pressed harder for Churchill. Halifax bowed out, perhaps

Sunday, 26 May 1940

understanding that he was not the man for the hour. At age 65, Churchill emerged as prime minister designate, his frustrating 'wilderness years' behind him and his lifetime ambition of leadership finally fulfilled.

When Churchill was leaving Buckingham Palace on the evening of 10 May after being officially appointed as prime minster by the king, his bodyguard, Inspector WH Thompson, congratulated him but added, 'I only wish the position had come your way in better times for you have an enormous task'. Churchill, tears welling in his eyes replied, 'God alone knows how great it is. I hope it is not too late. I am very much afraid that it is. We can only do our best'.

There was to be no honeymoon for the new prime minister. That very same Friday, 10 May, just before dawn, two million German soldiers marched into Belgium, the Netherlands and Luxembourg. To push back the Germans, five French armies and the British Expeditionary Force (BEF) crossed the French frontier and by nightfall were deep into Belgium. It was a perfect opening for Hitler, who danced a jig of joy upon hearing the news. With the Allies sucked into the Low Countries in the north and another two French Army Groups ensconced behind the 'impregnable' Maginot Line along the border with the Reich in the south, the Germans smashed through the 'impenetrable' Ardennes in the middle, effectively cutting the country in two and separating the two Allied wings from each other.

It was a new way of making war, the blitzkrieg, with the punch through provided by the mailed fist of concentrated armour. Fuelled by Romanian and Soviet oil, steely determination and methamphetamines, German panzers kept advancing day and night. No one had seen anything like it before. French and British defenders, despite their significant paper superiority in both men and materiel, including tanks and aircraft, were outwitted and outfought as the Wehrmacht columns struck towards the Channel ports in one sickle sweep and the road towards Paris in another.

Its fighting spirit broken well before any significant clash of arms, the French forces kept withdrawing on all fronts. In the north, the ten divisions of the BEF were forced to pull out as well, lest they get cut off alone and exposed inside Belgium. On 17 May, General Sir Edmund Ironside, Chief of the Imperial General Staff, wrote in his diary, 'At the moment it looks like the greatest military disaster in all history'.

Despite plans of a French-led counterattack, nothing materialised, and by the evening of Saturday, 25 May, General Lord Gort, commander of the BEF, decided that the best chance left for his troops was to make for the coast. Chased by German armour and harassed by the Luftwaffe, British troops had been doing this for days. Ironside, who earlier that week travelled to France to find ways to slow the German advance, could find none. 'We shall have lost all our trained soldiery by the next two days unless a miracle appears to help us', he recorded in his diary on 25 May. By then, some 400 000 Allied troops, including virtually all of

Sunday, 26 May 1940

Britain's trained army, were boxed in a shrinking pocket around the small resort town of Dunkirk.

The British got their first lucky break courtesy of their enemy. Two days earlier, on 23 May, the rampaging German divisions were ordered to stop. In hindsight, historians would consider Haltordnung, the 'halt order', to be one of Hitler's great mistakes that probably cost him the war. At the time, however, and despite the objections of panzer generals like Heinz Guderian, it made tactical sense. German infantry needed time to catch up to the panzers, days of heavy rain turned the approaches to the coast into muddy fields, and German intelligence badly miscounted the number of trapped Allied troops (they estimated fewer than 100 000, instead of the entire British army and then some), making their destruction seemingly less urgent.

Most importantly, Hermann Goering, Commander-in-Chief of the Luftwaffe, feeling overshadowed by the successes of the Wehrmacht, desperately needed an opportunity to shine. He convinced Hitler that his pilots alone could destroy the Dunkirk pocket, giving the panzer boys some well-deserved rest. While some ground pressure on the Dunkirk perimeter would continue, for the next few crucial days the greatest threat to the BEF would come from the German air force in the skies above them.

« • »

Some six weeks earlier, on 10 April, the day after German armies invaded Denmark and Norway, unknown to all but the attendees and the man who brought them together – Henry Tizard – a small meeting took place at the Royal Society headquarters in central London.

It was 2:30 pm, as five men sat around a large wooden table to discuss the contents of the Frisch–Peierls memorandum. If they were right and an atomic bomb could be built, the implications for the British government were staggering. A bomb like that, in British hands, could change it all. As it could in German hands, for worse.

Four out of the five gathered knew each other well; they were all Cavendish men, even if most had now moved on to other pastures. In addition to Oliphant, there was John Cockcroft, keeping his finger on the pulse of this latest development at the intersection of science and war, and Philip Moon, Oliphant's physicist colleague at Birmingham, as well as George Thomson of Imperial College at the University of London, who chaired and took minutes of the meeting.

The fifth man was the odd one out in a room full of some of Britain's most brilliant scientific minds, but the story he told those gathered only served to add urgency to an already momentous task. According to Jacques Allier, a well-connected French businessman and a spy, Germans were well aware of the possibility of an atomic bomb. Worse than that, the main source of 'heavy water', critical to building such a bomb, was now at risk of falling into Nazi hands. It was the Norsk Hydro power plant, at Rjukan in Norway, about a hundred kilometres west of the capital

Sunday, 26 May 1940

Oslo, now under German attack. Unknown even to Allier, a day earlier, the entire 185-kilogram stock of heavy water at the plant had been spirited away by the French secret service in collaboration with Norwegian authorities. But future production capacity remained intact. It was a race against time.

At its first meeting, the group, first dubbed the 'Thomson Committee', renamed itself the MAUD Committee. 'MAUD' originated from a telegram sent by Lise Meitner to her nephew Otto Frisch on 9 April shortly after Germany's invasion of Denmark to tell him that Niels Bohr and his wife were safe and concluding with the inexplicable last line 'TELL COCKCROFT AND MAUD RAY KENT'. Frisch passed the telegram on to George Thomson, and it was among the first items considered by the new committee. Deep and dark atomic secrets were sought to be divined from the mysterious reference. What sort of a code was this? Did it refer to a German bomb program? Sometimes the committee was even referred to as M.A.U.D. – with suggestions that MAUD stood for 'Ministry of Air: Uranium Development' – but the acronym was a nonsense and a clever disguise. Only years later did Bohr reveal that he had simply wanted the news passed to John Cockcroft and Bohr's former housekeeper, Maud Ray who lived in Kent. But the new name stuck. It was sufficiently nonsensical to throw nosy individuals off the scent and much better than 'Thomson Committee', whose chair was known to be researching uranium fission.

The two most urgent tasks for the new committee were to investigate the best way of separating uranium isotopes to obtain the highly fissionable uranium-235 and to clarify exactly how much uranium-235 was required for a 'critical mass' and therefore an explosion. With most scientists already fully committed to crucial war projects, resources were limited. All the attendees of the 10 April meeting would chip in whatever and whomever they could spare at their respective universities, but for now the bulk of work would fall on Birmingham, with Frisch to continue with his initial 'critical mass' calculations and Oliphant tapping Norman Haworth at the Chemistry Department to look into isotope separation to 'enrich' uranium.

It's not like Oliphant needed any distractions. The cavity magnetron was now undergoing final tweaks and would soon be independently tested in the field. With the fear of an imminent German invasion, it was not a moment too soon. This fear also motivated Oliphant to send his wife and children to the safety of Australia, a country at war but, unlike Britain, not in any immediate danger. He hated that they had to go but it would be one less worry on his mind. Rosa, Michael and Vivian sailed on 24 May.

《 • 》

On Saturday morning, 25 May, as Churchill and his generals despaired over the fate of the BEF, now relentlessly bombed and strafed by the Luftwaffe on the beaches and streets of Dunkirk, Florey was walking briskly across the blooming

Sunday, 26 May 1940

University Parks from his home to the laboratory. On that sunny morning in late spring, Florey was not distracted by Britain's dire struggle. His sole focus was the day's experiment, which would provide an answer for years of work and patient preparation.

Florey knew the consequences of success and failure. He was now 41, in early middle age. He was at the top of his game, widely respected throughout the, albeit small, world of pathologists and experimental physiologists. In the next 24 hours he was going to set in train events that might see him climb to the pre-eminence of Jenner, Pasteur and Lister in the pantheon of modern medicine – or fail miserably, having wasted precious time and resources chasing a medical chimera.

Having established that penicillin satisfied the first obligation of medicine of the Hippocratic Oath – 'Do no harm' – it was now time to determine whether it was effective inside living tissue: if it would kill deadly bacteria infecting an animal. Simple in concept, unpredictable in outcome. The strain and toxicity of the chosen bacteria, the quantity and purity of the penicillin to be injected, and the weight and health of the test mice would all have to be carefully accounted. Even then, all sorts of unexpected outcomes might emerge, rendering penicillin worthless.

Florey set Saturday, 25 May, as the day for what he termed the 'protective' mice experiment: to test whether penicillin protects against fatal infection in living animals. With the war virtually on Britain's doorstep, Florey and his team now frequently worked a day or two on the weekend. The genteel practices of prewar research had vanished amid the German blitzkrieg.

Just before 11 am Florey entered room 46 at the Dunn School. The scene was banal, perhaps even clichéd. It was a picture played out countless times in scientific experiments: eight albino mice, each in their 12-inch diameter circular glass cages, nibbling some crumbed biscuit on a saw-dust floor, and a somewhat bored lab assistant sitting on a chair in the corner. For James Kent it was just one of hundreds of experiments he had assisted so far under Florey's supervision and he was not yet aware of its significance. Nor were the mice.

At precisely 11 am, Florey proceeded to inject each mouse with a dose of about 100 million or so lethal streptococci bacteria. Having been so infected, four of the mice were placed back in their cage and left to their fate. This was the doomed 'control group'; they were certain to die from the virulent bacteria in their abdominal lining.

Of the other four – the 'protected' mice – two were treated an hour later with a single larger dose of penicillin and two with smaller doses given at intervals. The former pair received about one-fifth of 1 milligram of Heatley's brown penicillin powder, the latter pair about half as much initially, but this smaller dose was to be repeated several times throughout the afternoon and early evening.

For the moment, all Florey and Kent could do was wait. After years working together, they enjoyed a certain routine

Sunday, 26 May 1940

in such moments: Kent would brew tea on a Bunsen burner while Florey made notes detailing the experiment. As the mice twitched, scampered and nibbled, Florey and Kent ate their sandwiches looking for signs of distress. Heatley arrived later in the afternoon to join the vigil. By the time Florey let Kent go, at just after 6 pm, the behaviour of the mice gave little away. No conclusions could yet be drawn.

Soon Florey noticed a change. The four control mice, those not protected by penicillin, were starting to look sick. They were not eating, their heads had drooped, their eyes were closing and their breathing was laboured. This did not come as a surprise. Seven and a half hours after being infected with lethal bacteria, the mice were succumbing.

But what of the other mice 'protected' by penicillin? What happened next, or more precisely, what did not happen next, changed history. Florey recorded that the treated mice were not showing any distress. Indeed, three of the four were very well indeed, with one a little less lively. A little before 7 pm, with the first scent of success, Florey and Heatley went their respective ways to dinner.

Florey returned to the laboratory before 10 pm and noted the stark contrast in condition between the mice protected by penicillin and those not. One of the controls was 'nearly dead. The others in a poor way'. The other four mice were in 'good condition … cleaning themselves or eating'. Happy with progress so far but not wanting to spend the night in a small room with eight albino mice, four of them on their last legs, Florey went home.

Heatley arrived back at the lab in high spirits shortly after Florey had departed. He had unexpectedly run into a relative and enjoyed a lively evening repast complete with red wine at a local pub. Splashing himself with water to refresh, and perhaps inspired by the bottle of Mouton-Rothschild 1936, Heatley colourfully described the passing of the first hapless rodent:

> At three minutes to midnight the first of the four controls died. The creature staggered to its feet, lurched for a second or two and then fell down. It twitched once or twice and then died. Others very seedy.

By 1:30 am, two more of the control group had died. The fourth and final mouse hung on gamely for another two hours. At 3:28 am it too 'twitched several times and was dead'. For the last time a red cross was marked on the weekend's laboratory notes. While the control group had now all died from bacterial infection, the treated mice were apparently fine, with only one not looking completely well. The different administration of penicillin doses did not seem to affect the outcome. At 3:45 am, as darkness began to recede before the dawn, Heatley packed up, leaving the dead mice where they fell. He cycled home in 'relief, joy, happiness', with the first light of Sunday morning falling across Oxford.

《 • 》

Sunday, 26 May 1940

'Let Us Pray' declared the front page of the *Daily Express* on that solemn Sunday, 26 May.

Some 80 kilometres to the south-east of Oxford, in London, King George VI and Queen Elizabeth were indeed praying, hoping for a miracle. At 10 am, they were joined by Winston Churchill and high officials of the Empire in a National Day of Prayer at Westminster Abbey. The king and queen had arrived carrying gas masks and determined smiles. Millions across the nation prayed for peace and the safety of the British Expeditionary Force marooned in Dunkirk. Churchill left the service at Westminster Abbey early. He had much to do.

Former prime minister Neville Chamberlain described this Sunday as 'the blackest day of all'. Harold Nicolson, a junior minister in the government, and more optimistic than many, wrote to his wife, Vita Sackville-West, 'What a grim interlude in our lives! ... I am not in the least afraid of ... sudden and honourable death. What I dread is being tortured and humiliated'. Reflecting the general gloom, it rained for the first time in weeks.

On this grave Sunday, in addition to the high service in Westminster Abbey, there were three Cabinet meetings and the visit of a French delegation led by the Premier Paul Reynaud. Reynaud's message to Churchill was grim: France could go on no longer against Hitler's onslaught. Britain alone might have to face Germany and probably Italy. Churchill, a rotund man who enjoyed a healthy appetite and a good meal, 'ate and drank with evident distaste' that evening, according to his chief military assistant, General Hastings 'Pug' Ismay. As he rose from his table, he said, 'I feel physically sick'.

«•»

'It seems like a miracle!' Howard Florey was stunned. 'Of course it does. It *is* a miracle', replied the ebullient Chain.

With the country facing its greatest crisis in history – that Sunday was the precise midpoint in the five days from 24 to 28 May 1940 when 'Adolf Hitler came closest to winning the Second World War' – there were few other people in Britain who found a reason to be excited that grim and momentous day.

Four mice were dead, four mice lived, and the world would never be the same. Florey was huddled with Chain and Heatley in their small laboratory at the Dunn School. The usually imperturbable Florey could hardly speak. Chain, ever more excitable than his Australian colleague, started to dance. 'Chain was beside himself with excitement and Florey, as usual, was trying to conceal any outward sign of emotion', recalled Hugh Barry, another Australian Rhodes Scholar, present that day at the Dunn School.

But they had been lucky, unbelievably lucky, in ways they did not even know. Heatley's patient work and his unmatched talent for improvisation had managed to produce enough penicillin, of better quality than Fleming's, to make the experiment possible. But despite his best efforts, his seemingly 'pure' brown powder was actually more than

Sunday, 26 May 1940

99.5 per cent 'junk': sugars, fats, carbohydrates and proteins. If any of the hundreds of chemical compounds that made up nearly all of Heatley's powder had been toxic to mice or interfered with the working of penicillin, the experiment might have failed, with penicillin itself marked as the culprit. Yet again, it would have been put aside as an ultimately disappointing medical might-have-been. That too, Chain later wrote, 'border[ed] on the miraculous'.

For that matter, if instead of mice the testing was done on guinea pigs, the one species of rodent that does not respond to penicillin, the experiment would have been another dead end. The chances of something going wrong were immense. Despite all the creativity, hard work and professionalism of Florey and his team, success still required an enormous amount of luck.

Just one-fifth of 1 milligram of precious powder, itself less than half of 1 per cent pure – making the active antibacterial ingredient only about one-millionth of a gram in weight – diluted into 10 millilitres of distilled water and then injected into a mouse somehow did the trick. At that barely conceivable minute quantity, a 20-gram albino mouse was able to fight off over 100 million streptococci bacteria.

Florey was not entirely surprised by the results. Years of preparatory work and meticulous planning had made this outcome probable. But even so, the experiment laid bare not just penicillin's miraculous qualities but also the magnitude of the experimental and logistical challenges ahead.

That Sunday, Florey and Margaret Jennings set about planning further mice experiments to verify the initial findings. There would need to be many more trials to test infection with different bacteria and to develop the optimal size and frequency of doses.

And while the impact of the mice experiment on human health dawned on Florey and his team, no one could be certain until the success of clinical trials with hospital patients. Deflating the jubilant mood just a note, Florey reminded his team that an average human subject weighs 3000 times as much as a mouse.

So, the project required more penicillin. Much more. Heatley, whose principal task thus far had been to manufacture the substance, knew that current means of production were insufficient. With the painstakingly slow and laborious process as it now stood, human trials would be practically impossible. One miracle down, more required, lest all the hard work so far go to waste as yet another intriguing but ultimately unrewarding cul-de-sac in medical history.

《 • 》

At a few minutes before seven o'clock on that same Sunday night, the order was given to evacuate the British Expeditionary Forces encircled at Dunkirk.

The Royal Navy could not do it alone. For one, German bombing had destroyed much of Dunkirk's port, making it difficult for larger ships to pick up troops from the shore. Faced with a logistical nightmare, the Admiralty took on an

Sunday, 26 May 1940

idea floated by Churchill in the War Cabinet a few days earlier and appealed to the public and to yacht clubs for volunteers. In the end over 900 vessels – trawlers, minesweepers, fishing sloops, schooners, tugs, even the America's Cup challenger *Endeavour*, and a myriad of pleasure craft – all made the dash across the Channel. The most spectacular evacuation in military history was soon in full swing.

The Allies were trapped in a narrow pocket between the English Channel and the German Army, just 10 miles away. No one dared hope that any more than a small fraction of the force could be saved. Churchill thought that with a bit of luck, 30 000 troops might be rescued before the BEF was annihilated or forced to ignominiously surrender.

Just after 7 pm the first ship left Dover bound for the desperate men on the beach at Dunkirk. It was followed by a flotilla that did not let up until 1 June and continued in fits and spurts until 4 June.

Later, recalling Britain's resolve to fight on, Churchill summoned his penchant for high drama and fine rhetoric: 'There was a white glow, overpowering, sublime, which ran through our island from end to end'. It was bravado, in hindsight. In private, after a desperate visit to France at the end of May, Churchill said to Ismay, 'You and I will be dead in three months' time'.

But it was necessary bravado. The quick disintegration of the French army as an effective fighting force shocked Churchill just as it traumatised the French leaders and their nation. Even as 335 000 troops – more than ten times

Churchill's early estimate – were rescued from the beaches of Dunkirk by the navy and civilian flotilla and brought home despite constant German bombing, prospects were bleak. It was the miracle that Ironside had prayed for and it would become legend. But as Churchill himself said, 'Wars are not won by evacuations'. Brave Britons – well over a third of boats that took part in the rescue were destroyed or damaged – merely snatched a small, if much needed, victory from the jaws of Continental defeat.

《 • 》

Penicillin was no longer a laboratory curiosity; after that wondrous spring Sunday it became a potential life saver. Yet, enormous challenges remained. The Oxford team did not yet know the drug's chemical structure, so could not synthesise it. They did not know how it worked (only later would it be discovered that when bacteria, which multiply by division, split into two, penicillin stops the resulting gap in a cell wall from closing, allowing the surrounding liquid to rush inside and 'pop' them). But above all else, Florey desperately needed more of it: more penicillin to replicate the experiments and determine the ideal concentration and the best way to administer it, and much more to prepare for human trials.

Without successful human trials, there would be little interest from industry in producing the drug. Yet producing enough for use on a human would be a massive task without industry's help. It was a vicious circle. Florey had no choice

Sunday, 26 May 1940

but to turn the Dunn School laboratory into factory. He threw all the resources he could at Heatley and Chain with one command: make penicillin, and quickly.

Heatley tried all means to get more equipment to grow mould – using biscuit tins, dustbins, china plates, containers used for motor oil or sheep dip, even bedpans were pushed into service. Ever ingenious he also constructed extraction machines out of scavenged trash and leftovers; perhaps the world's most valuable substance gram for gram was being obtained using a cupboard-sized contraption worth £5 in materials. It was as improvised as the Dunkirk evacuation, and for the moment it had to do.

Florey was in a quandary. He was excited about the early results of his discovery but remained cautious. He had sworn his staff to secrecy not wanting to raise expectations publicly until the experiments had been replicated many times over. But to secure money and more staff from his academic backers he had to illustrate progress. Worse still, the actual production of penicillin – a smelly, messy and decidedly unglamorous activity, which would consume most of any new resources – was hardly an appealing facet to sell potential funders. The initial Rockefeller funding might have been a godsend for mice but it was not enough to take penicillin to the next stage.

Florey sought assistance from outside the university and in confidence approached the Wellcome Research Laboratories. But penicillin was not regarded as important for the war effort and the laboratories concentrated on work deemed of higher priority, like vaccines. It would be some time before the private sector with all its great industrial potential could be convinced to play a role in the project.

In the meantime, he gingerly approached Sir Edward Mellanby, head of the Medical Research Council, mentioning, 'We have been working with a substance that gives the greatest promise of being an important chemotherapeutic substance', but cautioning him, 'Naturally we do not wish anything to be said about this at the present stage'.

It wasn't much, but the Council awarded Florey's research enough to purchase much-needed equipment and either free existing colleagues from other less critical work or employ new staff to assist in producing penicillin. Only then, with Florey confident that momentum would be retained and production of penicillin scaled up, did he consent to publication of the results of the 'protective' mice experiments. In the summer of 1940, Florey chose to publish in the distinguished medical journal, *The Lancet*.

《 • 》

The Luftwaffe did not succeed in annihilating the BEF in the Dunkirk pocket. The bulk of the British Army had managed to make its way back home, even if forced to leave their guns, artillery, tanks and ammunition behind. It was a great loss but not a complete disaster. Materiel and equipment could be replaced in time; trained troops took a lot longer.

Sunday, 26 May 1940

As the Dunkirk evacuation was winding down, Churchill received intelligence from decrypted Wehrmacht signals that the Germans were seeking to complete their invasion of France before turning all of the Third Reich's resources against Britain. This gave Churchill some breathing space and time to prepare, but not as much as he might have hoped.

On 13 June, Paris, in order to prevent its destruction, was declared an open city as the French Government fled to Bordeaux. Marshal Phillipe Petain, France's famed First World War hero, was recruited to rescue his nation and airily declared that France would make no shameful surrender. He then promptly sought an armistice with Germany. It was signed on 22 June in the same railway carriage that Germany had surrendered to France and the Allies to end the First World War, an acid touch of humiliation to go with the French rout. Britain was now alone.

'"The Battle of France" is over', Churchill summed up the perilous predicament, 'I expect the Battle of Britain is about to begin'. Just six weeks after Churchill became prime minister, German armies stood poised to attack across the English Channel. An elated Fuhrer arriving in defeated France could not resist a sojourn to the coast and to peer through a view finder at the white cliffs of Dover. Nazi soldiers camping on French beaches sang 'We're sailing against England'.

No one knew exactly when, but it was expected to be soon. The government contemplated sending the king and queen to the safety of Canada, but the royals refused to go. Upon learning this, Churchill sent a Tommy gun to the king. The queen learned how to fire a revolver, telling Harold Nicolson, 'I shall not go down like the others'. Even the princesses, Elizabeth and Margaret, joined in with firearms training in the grounds of Buckingham Palace. As the royals practised, Churchill and the military command scrambled to prepare for the German onslaught. It was to be, as Churchill promised, 'their finest hour'. And it was also to be his.

《 • 》

For most Britons the summer of 1940 was one of 'terrible beauty', with magnificent weather almost welcoming a German invasion. 'They expect an invasion this weekend', wrote Harold Nicolson in his diary on 11 July. He could have written it almost any day throughout July and August.

Yet while the German High Command was planning and preparing troop landings, codenamed 'Sea Lion', Hitler made no final decision. Maybe it would not be necessary. Germany's overarching interest, after all, lay in the east and the Lebensraum, or the living space, it offered for the German master race. What was needed in the west was acquiescence, however secured. Hitler would not necessarily need to bring Britain under his rule if he could bring it to the negotiating table, but to do so he would first need to bring it to its knees.

To succeed, the Nazi war machine would have to sufficiently degrade British defensive capability and fighting spirit. Destroying the RAF and other military assets might be enough to either force Churchill's hand or, better still, force Churchill out of office in favour of a more sensible leader. Goering knew that his superb aircraft outnumbered those of the RAF by two to one. Pumped up by the strong performance of the Luftwaffe in the Battle for France (if forgetting the failure to destroy the British Army in Dunkirk), the *Reichsmarschall* assured the Fuhrer that it would take only a few days, a couple of weeks at most, to mop up the British, saving the Wehrmacht the blood and the effort.

The Luftwaffe commenced its onslaught on 10 July. At first, the Germans attacked ports, shipping and radar stations, and then in its most devastating tactic, targeted RAF airfields and installations. This served a double purpose: even if the destruction of their air force did not convince the British to sue for peace, it was a necessary precondition for any successful cross-Channel attack.

Then, on Saturday, 24 August 1940, the Germans dropped their bombs on the City of London, its financial centre, and the working-class residential areas of the East End. This was the first time the Germans had bombed London since the Gothas in 1918. The pilots most likely were off course and confused about their location, but for Churchill and Londoners this was an unprecedented, deliberate and callous attack on non-combatants. An enraged Churchill ordered the RAF to bomb Berlin; in retaliation, Hitler instructed his Luftwaffe to start bombing

British cities without regard for civilians. The Blitz now escalated in intensity and misery – most likely because of a tragic navigational error.

《 • 》

Earlier on that same Saturday, the respected British weekly medical journal, *The Lancet*, its London headquarters only a short distance from the soon to be unleashed inferno, published an article with a non-descript, clinical title 'Penicillin as a chemotherapeutic agent'. About two pages long and written in terse language, this pithy paper signalled a revolution in humankind's battle to conquer infection and disease. As so often with scientific writing, the dryness of the prose was in reverse proportion to the startling nature of the discovery. The piece, co-authored (in alphabetical order) by Chain, Florey, Gardner, Heatley, Orr-Ewing and Sanders, begins innocuously enough:

> In recent years interest in chemotherapeutic effects has been almost exclusively focused on the sulphonamides and their derivatives. There are, however, other possibilities, notably those connected with naturally occurring substances. It has been known for a long time that a number of bacteria and moulds inhibit the growth of pathogenic microorganisms. Little, however, has been done to purify or to determine the properties of any of these substances.

Sunday, 26 May 1940

The paper tactfully acknowledges the earlier work of Fleming and others but reminds readers, in the politest of terms, that these pioneers were unable to isolate penicillin. Florey's team, however, did it:

> During the last year methods have been devised here, for obtaining a considerable yield of penicillin, and for rapid assay of its inhibitory power. From the culture medium a brown powder has been obtained which is freely soluble in water. It and its solution are stable for a considerable time and though it is not a pure substance, its anti-bacterial activity is very great.

The mice trials are then recounted, and a useful graph compares the fate of mice infected with bacteria and the effect of penicillin on mortality rates. 'The results are clear cut', the paper concludes, 'it would seem a reasonable hope that all organisms inhibited in high dilution *in vitro* will be found to be dealt with *in vivo*'. The Latinate opaqueness of the last sentence cannot disguise the staggering nature of the findings: penicillin had proven its potential to defeat different lethal bacteria in test tubes and also destroy at least some bacteria in test mice.

In a typically cautious and measured comment on the paper, *The Lancet* editors remarked, 'What its chemical nature is, and how its acts, and whether it can be prepared on a commercial scale, are problems to which the Oxford pathologists are doubtless addressing themselves'. All that, as well as the ultimate test, of course: will it work on human beings as well as mice?

The editors needn't have worried. These questions would now preoccupy the team in their every waking moment. Florey himself would carry the additional burden, unmentioned in the editorial but perhaps the heaviest of all, of raising money from funding bodies. But he had high hopes. Initial secrecy now abandoned, Florey was expecting *The Lancet* paper to generate significant interest in penicillin and its miraculous powers; interest that would finally help to overcome the scientific and technological roadblocks slowing down his under-resourced and overstretched team.

He was wrong. Mired in the immediate needs of war production and still unwilling to invest in a speculative drug, no pharmaceutical companies came forward. Florey and his team would somehow have to find a way to produce penicillin for the forthcoming human trials themselves.

Florey was even more surprised that few scientists showed any interest. Perhaps, he reasoned, they did not understand the implications of what his team had discovered. Or, more likely, they were yet to be convinced that penicillin could be mass-produced from mould or synthesised. Whatever the reason, Florey's findings, charged with possibility and hope for the fight against infection, did not immediately rouse the scientific community.

There was, however, one notable exception: Alexander Fleming. One can only imagine the Scot's thoughts when he

Sunday, 26 May 1940

flicked through his 24 August copy of *The Lancet* to find an article not only confirming the antibiotic powers of penicillin but forecasting the possibility of a miraculous drug for use in humans suffering infection. He did not wait around. He phoned Florey on Sunday, 1 September, and arrived the next morning to a quiet Oxford from a London already growing accustomed to the Blitz. When Florey mentioned to Chain that Fleming was about to visit, Chain was stunned. 'Good God!' he replied, 'I thought he was dead'.

Fleming, elfin and inscrutable, his kerchief adjusted and bow tie neatly in place, arrived at the Dunn School and greeted Florey at the door: 'I hear you've been doing things with my old penicillin. I'd be interested to look around'. Much has been made of these few words in the years since. Those more suspicious argued that Fleming, with an eye on future accolades, was trying to establish precedence for his work conducted more than a decade before. It is true, he had abandoned penicillin in favour of research on vaccines and later sulphonamides, and had not mentioned penicillin for the better part of ten years. But he later insisted that he had never lost faith in penicillin's healing properties, even if having discovered the substance he could not reproduce it.

Whatever the case, Florey and his colleagues had no qualms or suspicions about their surprise guest. Nothing was held back from Fleming in his two-hour tour of the facilities, with Heatley giving Fleming a full brief of the team's latest methods of producing penicillin. It was a strange visit. For someone who understood the significance of the antibacterial qualities of penicillin, Fleming asked no questions and departed the Oxford laboratories without a hint of praise for the work they had done. He was never to return. Chain was not convinced that Fleming even understood the difficulties Florey's team had overcome or the full ramifications of their work. Whatever Fleming's true feelings, the slightly testy atmosphere of his visit was but a mild precursor to later difficulties in his relationship with Florey.

《 • 》

At the height of the Battle of Britain in mid-September, Churchill visited RAF Fighter Group 11 Headquarters at Uxbridge, a few miles west of London. He asked Air Vice Marshal Sir Keith Park, the commander responsible for the fighter defence of London and South East England, 'What reserves have we?' Park replied, 'There are none'. While about 115 new Hurricanes and Spitfires were coming off the production line each week, the Germans were shooting them down faster than they could be replaced. The RAF, in turn, was downing even more enemy planes. This could not go on forever; eventually one side or the other would run out of aircraft.

Much has been written about the reasons for the RAF's victory in the Battle of Britain. On 20 August, with the full fury of the Luftwaffe's assault not yet spent, Churchill famously credited the courage and ingenuity of the young British – and Polish and Czech – pilots, 'Never in the field

Sunday, 26 May 1940

of human conflict was so much owed by so many to so few'. But, above all, it was long-wave radar that enabled Britain to wear the Luftwaffe down.

Unsightly but reliable, the Chain Home radar towers were able to detect enemy airplanes up to 200 kilometres away. Information from the radar stations was passed via the Fighter Command to the closest squadron base which, in turn, ordered its pilots into the air. This procedure, dubbed the 'scramble', took 15 minutes and it was just enough time. Out of the clear blue sky, to the shock of German pilots, a haze of Spitfires and Hurricanes would emerge to attack their bombers and fighter escorts. How, the Germans wondered, did the RAF fighters know where to find us?

If the Germans had managed to knock out enough radar stations to degrade and compromise the system, they would have blinded the desperately overstretched RAF. Their failure to do so, indeed the failure of German intelligence to comprehend the advantage that Chain Home was giving to the RAF, was crucial to Germany's inability to win control of British skies. So ingenious was this radar defence system that it has been described as 'one of the greatest combined feats of science, engineering, and organisation in the annals of human achievement'. The early faith of Dowding and Watson-Watt in long-wave radar technology was vindicated when it counted, in the heat of battle.

Yet Chain Home, and British civilians, soon became a victim of this success. The Battle of Britain, in its first phase, from July to August, was a daytime battle for air superiority, a battle that the British won – just. But unsustainable losses forced the Luftwaffe to commence night-time operations.

German bombers now began to fly without fear and bomb with impunity – subject to the occasional spray of anti-aircraft fire. While coastal radar stations of the Chain Home network might pick up the intruding enemy bombers, the scrambled RAF fighter pilots simply could not see them, unless it was a clear night with a full moon. Pilots were advised to eat more carrots to sharpen their eyesight. Some imaginative planners even recommended recruiting cats into cockpits; with their legendary night vision they would presumably meow in the direction of the approaching enemy. It was of course fanciful but born of desperation. What the RAF pilots needed was not carrots or cats but their own onboard radar to give them the eyes that could pierce the darkness and guide them straight to the enemy.

What the German bombers gained in safety they lost in accuracy: flying in the dark made locating targets much more difficult. It also likely cost Hitler victory in the Battle of Britain. The mistaken bombing of London on 24 August led to a spiral of tit-for-tat retaliations against civilians. But in changing strategy from destroying the RAF – which it came very close to achieving – to destroying British cities, Germany failed to crush Britain's air defences enough to force an armistice – or prepare for a successful invasion. It was a 'critical mistake', and while it certainly did not seem so at the time, it was arguably Germany's first defeat.

Sunday, 26 May 1940

As the history of the next few years would show, the results of air offensives, while costly to both sides, did not affect the outcome of the war. In neither Britain nor Germany were civilians so traumatised as to urge their political masters to parley with the enemy (in totalitarian Germany this would have been unlikely in any case). Baldwin and others were wrong. The indiscriminate bombing of cities may even have hardened resistance with the 'spirit of the Blitz' becoming an enduring British rallying cry. If the bombing made civilian and soldier alike share in the experience of total war, they also shared the determination to fight on.

《 • 》

Long-wave radar, as revolutionary as it was, could not win the war. At best, it might stop Britain from losing it. Someone else would need to pick up the torch from Watson-Watt and chart radar's next chapter in war.

The industrial areas in Birmingham were first bombed in early August, but the university was left alone until late October 1940. During the Birmingham Blitz, which lasted over 33 months, Oliphant's laboratory was hit two or three times. Incendiary bombs even penetrated the roof of the new Nuffield Laboratory – home of Oliphant's prized but for now mothballed cyclotron.

Around that time Oliphant got to know well Florey's old friend, Keith Hancock, who was now a professional colleague at Birmingham. Dean of History by day, Hancock donned the uniform of a sergeant of the Home Guard by night. 'More than once', Hancock later recalled, 'I had to report lights being shown from Oliphant's lab ... I didn't know what he was working on'. Meanwhile Oliphant served as a fire-spotter scanning the horizon for flames from the elevated portholes of the old Poynting Physics building and then later grabbing fitful sleep in his office nearby. He got lucky. He arrived back at his office one morning to find an unwelcome guest: an unexploded 500-pound bomb, a present from the Luftwaffe, behind a bookcase. The army soon defused the bomb with a grateful Oliphant left counting his blessings. He himself remained largely unscathed, except for walking into a glass door half-asleep during an air-raid alert.

The cavity magnetron prototype was first tested in the field back in June, just as France slouched towards ignominious defeat. What the boffins at the radar development station at Swanage on the Dorset coast saw was new and remarkable. They first picked up radar echoes from tinplate hung up on the nearby Downs and were soon able to receive echoes off boys on bicycles nearby.

By early August, with the war now raging above Britain, the magnetron had advanced to churn out 12 to 15 kilowatts, better than 30 times the output of Randall and Boot's primitive prototype only six months before, and at least 1000 times the output of the klystron or any other alternative. Members of Oliphant's Birmingham team were already able to accurately locate the Isle of Man at night as well as orient themselves safely during morning fog. On

Sunday, 26 May 1940

12 August, microwaves generated by the cavity magnetron tracked a lone aircraft several miles away moving down the coast. It was the first time that a prototype short-wave radar had followed an aircraft in flight. The dream of pinpointing aircraft had come a step closer.

7

Alone in the storm

The blasted black box was the bane of his life.

He couldn't quite understand why, at 29 and the youngest member of the mission, he was tasked as its keeper. The responsibility weighed on him more than the metal box itself. There was nothing remarkable about it on sight – just your standard sturdy solicitor's deed box – but its contents could change the course of war.

Eddie 'Taffy' Bowen had tried to put the box into a safe at the Cumberland Hotel, his overnight accommodation in London, but it didn't fit. Instead, he slid it under his bed and spent a restless night on top, guarding it with his body. He now sought to squeeze the box into the back of a taxi, but the driver wasn't having it. The box would have to travel on top of the car. With a train to catch and no time to argue, Bowen looked on helplessly as the stroppy cabbie strapped the box to the roof and took off for Euston railway station.

Over ten anxious minutes through Marylebone's morning traffic, Bowen contemplated his fate, and Britain's, if the ropes should give way and the box's contents were damaged, lost or, even worse, fell into the wrong hands.

For inside it were manuals, diagrams and blueprints of technical innovations that would revolutionise warfare: the jet engine, the proximity fuze, plastic explosives, self-sealing fuel tanks, improved gunsights and rocket designs. But the two most significant were fresh from Oliphant's lab: the plans for an atomic bomb, evolved from the Frisch–Peierls memorandum, and – the only physical object in the box – a palm-sized prototype 'Cavity Magnetron Number 12'; the 'precious magnetron', as Cockcroft called it. The 'briefcase that changed the world' was starting its journey to the United States.

Bowen breathed a sigh of relief as the taxi arrived at Euston Station. But his anxiety quickly returned when, struggling with his other luggage, an officious porter grabbed the black box, put it on his shoulder and started off to find Bowen's seat on the 8:30 am train to Liverpool. For a second, he lost sight of the porter, but he spied the black box and followed it, as it weaved above the heads on the crowded platform.

Bowen raced toward the first-class carriages to his reserved seat. The blinds of his compartment were drawn and large notices posted on both windows warned others off. The precious black box was already on the luggage rack. Just before the train was to leave, a smartly dressed man turned up outside the door. For a second, Bowen was worried. When he discovered the man had been sent to protect him and his cargo, he uttered a sigh of relief.

The train pulled into the Liverpool docks in the early afternoon of 29 August 1940. The city had suffered its first major air raid, attacked by 160 German bombers just hours earlier. Fortunately, neither the railways nor the docks were damaged. As Bowen's ship, the *Duchess of Richmond*, left its mooring under cover of darkness that night, bound for Halifax in Newfoundland, German bombs started to rain down again. The *Duchess* had to halt and seek refuge by the shore.

The liner finally departed in the morning. Fast enough to outrun *Kriegsmarine* surface vessels, it did not need an armed escort. U-boats were another matter. The ever-present threat of prowling submarines with their deadly torpedoes meant the *Duchess* had to zig-zag across the Atlantic. In the event of an attack and the likelihood of the vessel being lost, the captain's instructions were to throw the black box overboard to keep it out of enemy hands.

《 • 》

So began the most important exchange of scientific information of the Second World War. The objective of the British Technical and Scientific Mission, or the Tizard Mission, was to spearhead vital cooperation with the United States. Born of Britain's desperate search for friends and assistance in the summer of 1940, the mission went on to establish formal protocols and informal trust between the two English-speaking powers. The key to building that trust was sharing the fruits of British military and scientific research with the less advanced American side. But it almost didn't happen.

As the *Duchess of Richmond* set sail in late August 1940, Britain had been at war for almost a year. It was not going well. France's inglorious surrender two months earlier had left Britain with no allies in Europe. Forced to look to its far-flung Empire and dominions for help, scarce overseas supplies were now at the mercy of the German naval gauntlet. While the RAF had so far managed to hold on – barely – in the opening jousts of the Battle of Britain, the Luftwaffe went on to unleash a concerted bombing campaign against cities in a run up to an expected invasion. The outlook for the defenders was grim. The British Army, miraculously evacuated, had to leave virtually all its materiel behind at Dunkirk. It needed to be replaced.

Oliphant's lab had thrown Britain a lifeline. In fact, Britain had many good ideas but neither the time nor the resources to develop them on a scale that would make a difference. Yet there was no consensus among scientists, mandarins and politicians as to whether the United States could be entrusted with secret British technologies and relied on to mass-produce them for a war that they were still staunchly refusing to enter. Oliphant himself had no doubts what should be done, writing to his friend James Chadwick in June 1940 that 'if things go really badly with this country there is a great deal to be said for investigating any possibility which offers a chance of hitting back from the New World'.

After months of to-ing and fro-ing, Churchill finally made up his mind. For generations, American heiresses – including his own mother – travelled to Britain to match their wealth and glamour with the prestige and history of England's titled aristocracy. Now the tables were turned, and Britain voyaged to America to woo and dazzle it with science. The tantalising secrets like the cavity magnetron and other gems would be offered like a rose on a first date.

There were signs that transatlantic relations were moving in Britain's favour. The United States had just agreed to a 'destroyers for bases' deal, lending Britain 50 First World War destroyer ships to protect convoys, in exchange for the right to use British naval and air bases in Newfoundland and the Caribbean. So, as the *Duchess of Richmond* steamed across the Atlantic, it was carrying, in addition to half a dozen members of the Tizard Mission (Tizard himself had flown to North America two weeks earlier), a thousand men of the Royal Navy who would crew the refurbished destroyers back to Britain.

《 • 》

A few months earlier, on 12 June 1940, President Roosevelt had met with the President of the Carnegie Institution for Science, Dr Vannevar Bush. Dark haired and angular, with the air of a New England patrician and the look of a hungry lawyer, Bush was dubbed the 'science czar'. Well-connected in the worlds of both politics and science, Bush was stirred to action by Nazi aggression in Europe. He wanted the president to set up a body that would bridge the gap between science and the military. It would identify new technologies with military application and fund their research and development. After a short 15-minute meeting in the Oval Office, Bush handed his brief to the president who scrawled on it 'O.K., FDR'.

With his signature, Roosevelt kicked-started American scientific and military cooperation that bore the seeds of triumph in the Second World War. 'The Engineers of Victory' went on to create a cornucopia of deadly technologies, many still in use today. At the centre of this development was the National Defense Research Committee (NDRC) and its successor, the Office of Scientific Research and Development (OSRD), both chaired by Bush.

Tizard instantly hit it off with Bush who, like Tizard, was a supreme technocrat – cool, solicitous and bright. Cockcroft, Tizard's deputy on the mission, meanwhile, shamelessly leveraged his longstanding friendships with Alfred Loomis, head of the NDRC's Microwave Committee, and Ernest Lawrence, Nobel laureate, inventor of the cyclotron, perhaps America's best-connected physicist. Yet charm and personal rapport could only go so far. The Americans, convinced of their own scientific superiority, were sceptical and suspicious of British motives. Trust took time to build.

In the end, it was the 'Show-and-Tell' that did it. There is a touch of a travelling carnival in the accounts of Bowen and

Cockcroft: at opportune moments, Tizard's wizards would pluck an item from the magic black box to gasps and cries from the attendant servicemen and scientists. The road to cooperation ran through a sideshow alley.

The Americans were well aware of the profound military advantages of microwave radar. They had developed a transmitter capable of emitting 10-centimetre waves, but it lacked the necessary power and their research effort had reached a dead end. By the end of the summer of 1940, the Loomis' Microwave Committee was ready to write a report, 'a sign', according to one of its members, 'that we didn't know what to do next'.

Bowen and his colleagues had been hinting at the existence of the cavity magnetron from the moment they stepped off the *Duchess*. But the magic only started to happen when he and the rest of Tizard's mission met with senior American military officers and civilian physicists at Loomis' New York City apartment. On 19 September, after a big dinner, with at least one admiral 'conspicuously drunk', the contingent retired to a large room. When Bowen quietly produced the magnetron, the Americans were 'shaken' to learn of its power – though not quite stirred into action.

The 'crucial' meeting took place at Loomis' home in Tuxedo Park, about 80 kilometres north of New York City, on the weekend of 28 and 29 September.

Alfred Loomis was rich, suave and discreet, an establishment lawyer who had made a fortune in investment banking. Possessing a disarming smile and, according to a less generous colleague, a 'devious charm', he was no dilettante. His estate housed a laboratory where he undertook pioneering research with equipment that many universities could not afford. Loomis understood science and was intimately connected to politics and power. He got things done. This is why he was tapped to work on the NDRC. Bowen liked him immediately.

Whisked in a Loomis limousine through New York state's golden autumn foliage, Bowen and Cockcroft were delivered to the gated community of Tuxedo Park, a hunting and fishing refuge founded for Gilded Age high society. Bowen was agog – it was a very different world from the one he had grown up in. His birthplace, Cockett in south Wales, would never lend its name to formal attire.

After a weekend of entertainment and briefings about US radar research, the stage was finally set for the black box. On Sunday evening, joined by some of America's most influential radar scientists and advocates (including Ernest Lawrence), Bowen pulled out Britain's prize exhibit, the cavity magnetron. He explained that this palm-sized contraption of metal discs, glass tubes and copper wiring 'increased the power available to US technicians by a factor of 1000'. With the cavity magnetron, a ship or an aircraft could now spot the conning tower of a surfaced submarine at night. Bowen later recalled, 'The atmosphere was electric – they found it hard to believe that such a small device could produce so much power and that what lay on the table in front of us might prove to be the salvation of the Allied cause'.

No one was more excited than Loomis. The magnetron was the answer to the Microwave Committee's prayers. He would now run with it.

Bowen recalled that by the time he and Cockcroft left Tuxedo Park the next morning, it had been agreed that Loomis' committee would sponsor the Bell Telephone Company to manufacture a small batch of the cavity magnetrons. Given that there were only 12 magnetrons in existence, Bowen and Cockcroft were overjoyed and soon passed on the good news to Henry Tizard, by then back in Britain. Finally, tangible success.

Loomis, already pro-British in sympathies, had now become an apostle for scientific and military cooperation, preaching the good news in high places. One of Loomis' early converts was Henry Stimson, Roosevelt's respected Secretary of War, who recorded in his diary on 2 October 1940:

> Alfred Loomis came in in the afternoon, full of excitement over his interviews with the British and with the scientists, and he was full of the benefits that we were getting out of the frank disclosure by the British to us of their inventions and discoveries of methods they have made since the war. He said we were getting the chance to start now two years ahead of where we were.

Now powerfully backed by Ernest Lawrence, Loomis followed through on his promise, getting prototypes quickly built by Bell. Not wasting a minute, he organised government funding to further develop and improve the device at the Massachusetts Institute of Technology's Radiation Laboratory, or as it soon became commonly known, the Rad Lab, which was set up specifically for this purpose. Watson-Watt deservedly called Lawrence the Rad Lab's 'godfather'. 'No one else in American physics', he wrote, 'could so effectively have persuaded the brilliant young physicists of his country to put *their* shirts in turn on the dark horse from Oliphant's Birmingham stable – without being told to what end they were being asked to dedicate "the best years of their lives"'.

'If Lawrence was interested in the program, that was what I wanted to be in', said Lee DuBridge, the Lab's first director, speaking on behalf of his scientific generation.

It was at the Rad Lab that future Nobel laureates Isidor Rabi, Norman Ramsey and Luis Alvarez first encountered the genius of the British invention, Alvarez recording, 'it was a remarkable invention ... [w]e were ... awed by the cavity magnetron. Suddenly it was clear that microwave radar was there for the asking'. Not yet 30, Alvarez would later witness the bombings of Hiroshima and Nagasaki from an observation aircraft *The Great Artiste* and win the Nobel prize for his contribution to particle physics. His very first entry in his Rad Lab notebook references a 'model to find resonant frequencies of the O. Tube'. The abbreviation stands for 'the Oliphant tube', as some Americans named the cavity magnetron.

By the early spring of 1941, the Rad Lab delivered prototypes with wavelengths of just 3 centimetres, making for clearer vision, less interference and greater lethality. Airborne radar was field tested successfully in March 1941 and by the middle of the year the Rad Lab had even adapted the technology for anti-aircraft guns, making them significantly more accurate. Oliphant's baby was now in good hands.

《 • 》

The Tizard Mission's achievement in progressing radar technology in America was not replicated with the other great project from Oliphant's stable.

President Roosevelt was first made aware of the prospect of an atomic bomb in October 1939, in a letter from Albert Einstein (actually written by three Hungarian physicists more knowledgeable about the issue, Leo Szilard, Edward Teller and Eugene Wigner). The missive warned Roosevelt that Germany might be developing atomic weapons and suggested that the United States should do likewise, so as not to be left defenceless.

After an intermediary, economist and personal friend Alexander Sachs presented the letter to him, Roosevelt reflected, 'what you are after is to see that the Nazis don't blow us up'. 'This requires action', he concluded. As is customary in politics, that action was forming a committee.

The Advisory Committee on Uranium was chaired by Dr Lyman Briggs, a government soil scientist and now the Director of the National Bureau of Standards. Briggs, slow and ponderous, was ill-equipped to inquire into the possibilities of a revolutionary new technology with earth-shattering implications for war and peace. Other participants were from army and navy ordnance departments and had little expertise in nuclear physics.

After ten days of perfunctory pulse-taking in the American physics community, the Uranium Committee reported back on 1 November 1939 on the complex possibilities of nuclear fission. Military applications were purely speculative, they concluded. With that, everyone calmed down and did little. Briggs' committee received a princely sum of $6000 to pursue further inquiries and a few months later, in June 1940, it was incorporated into Vannevar Bush's newly created NDRC. This administrative card shuffling made little difference to the American atomic effort, which remained largely moribund well into the second half of 1941.

Yet the British had made a concerted effort to share their progress in atomic research with the American side. One of the first actions of the MAUD Committee after it formed in April 1940 was to ask one of Tizard's friends, Professor AV Hill, to inquire into the state of nuclear science in the United States. Hill, a Nobel laureate and Member of Parliament for Cambridge University – a unique double distinction – was at that time posted to the British Embassy in Washington to promote war research. Unburdened by the secrecy surrounding radar and other top-secret military projects, American scientists spoke freely to Hill. A strong chorus told him that there was 'no possibility within practical range' of a uranium bomb; it was a 'wild goose chase' and a 'sheer waste of time'. The consensus among science administrators like Bush and Loomis, echoing the conclusions of Briggs' committee, dubbed the prospect a pie in the sky.

In July 1940, when Churchill made the decision to share Britain's scientific secrets with the United States, Professor Ralph Fowler passed on the Frisch–Peierls memorandum and other early information detailing the feasibility of an atomic bomb. Fowler, Britain's well-connected scientific attaché in Ottawa and shortly in Washington DC as well, was a scientist of note – one historian described him as a 'man whose experience combined an almost unique mixture of expert knowledge and diplomatic skill' – and the son-in-law of the late Ernest Rutherford. Fowler would remain the official conduit for all atom bomb information to the Americans until the summer of 1941, by which time he was joined and assisted by the new head of the British Central Scientific Office in Washington, Dr Charles Darwin, grandson of the great naturalist. Science, like business, often ran in families.

Truth be told, while they carried updates on the Frisch–Peierls memorandum, the members of the Tizard Mission considered radar more important and much more likely to influence military outcomes. Accordingly, there was only one meeting of any consequence about nuclear research during the Mission's two-month stay in America and this touched only on techniques to separate the uranium-235 isotope from uranium-238 and the safety of international uranium stocks.

Tizard realised neither how topical the latter discussion was nor how crucial his own role in the saga. From early spring 1939, with public discussion of a potential 'super bomb' or 'uranium device', British scientists started to worry about the supply of uranium – both to Britain and Germany. Thomson was the first to sound the alarm and Tizard soon acted. On 10 May 1939, in London, Tizard met with Edgar Sengier, a Belgian director of the Union Miniere du Haut Katanga, the owners of the richest uranium ore deposits in the world. 'Be careful', he told Sangier, 'and never forget that you have in your hands something which may mean a catastrophe to your country and mine if this material were to fall in the hands of a possible enemy'. This message was bluntly reinforced by French physicist Frederic Joliot-Curie and his colleagues, but Sengier's plan to ship the ore to Joliot-Curie came unstuck with the outbreak of war.

A few months later, Sengier fled to New York, from where he continued to direct Union Miniere's operations. In September 1940, in an act of great foresight and fearing German invasion of the Belgian Congo, Sengier had much of the mined uranium ore (around 1200 tons) discreetly shipped to the United States. There it lay largely unknown and undisturbed in a Staten Island warehouse until two years later when, amid a desperate search for high-grade uranium ore to fuel the Manhattan Project, as the American program to build the atomic bomb became known, someone recalled the Belgian cargo. It was Sengier's ore that made American nuclear weapons possible.

All that, however, was far in the future. At the time of the Tizard Mission, Cockcroft, though still sceptical that an atom bomb could be built in the short term, thought 'that the uranium investigation in North America deserved a more vigorous attack than it was receiving'. Rather optimistically, however, he considered it was only a few months behind Britain. Fowler was more realistic, reporting back to Hill that while the Americans talked a big game they 'have damned little to offer'.

They also had increasingly fewer excuses. Following the Tizard Mission, the United States and Britain agreed that military–scientific cooperation should be formalised and heightened. As a result, the MAUD Committee's progressive reports and minutes were now routinely sent to the United States for use by American scientists. On a couple of occasions, representatives of Bush's NDRC, including Harvard physicist Kenneth Bainbridge, even attended meetings of the MAUD Committee in London.

Bush read Bainbridge's reports and they are credited with convincing him to change his mind and support an active American atomic research program by early 1941. But if Bush was an early convert to the possibility of a bomb, he did little to indicate his new-found faith. While the official US history records that by July the MAUD Committee had influenced Briggs to recommend the possibility of a chain reaction be investigated, not a single member of Briggs' committee really believed that uranium fission would be of critical importance during the Second World War.

In short, neither the Frisch–Peierls memorandum nor subsequent information from the MAUD Committee provided by the Tizard Mission, Hill, Fowler, Darwin and others made the Americans sit up and take note. This can be partly explained by the different emphases placed on uranium fission on both sides of the Atlantic. For the Americans, lulled by the comforts of neutrality, a bomb was a distant and marginal prospect, with peaceful application of nuclear energy carrying far greater promise in the short term. The British, by contrast, facing an existential threat from Nazi Germany, singularly concentrated on the prospect of a new game-changing super-weapon.

Thanks to Frisch and Peierls, the bomb was already in rough outline when the MAUD Committee first met in April 1940. But their physics and their assumptions had to be double-checked and confirmed. Moreover, the logistical and practical problems were immense. The only point of agreement was the possibility that Germany's famed physicists might be pursuing a bomb, which made it urgent to proceed. If a bomb was possible, Britain had to build it first. It would take more than a year of detailed work, but Frisch and Peierls' original insights were progressively confirmed.

Despite an initial reluctance to have the two 'enemy aliens' involved in this top-secret project, Thomson soon relented, convinced by Frisch and Peierls, that they 'had thought a great deal about the problems already and might well know the answers to important questions'. Such expertise should not go to waste, so official rules were bent and the two were soon engaged in work on MAUD's technical sub-committees.

At their first meeting Peierls warned Thomson that one of the greatest technical challenges was separating the uranium-235 isotope from the far more abundant uranium-238. Peierls suggested that the respected Professor Franz Simon of Oxford's Clarendon Laboratory be approached to lead the isotope separation work. As Simon, born in Berlin and of Jewish faith, had only been naturalised a year before, the same issues arose that Peierls had faced. And once again, while the committee continued to hesitate, Peierls, with Oliphant's nod, pre-empted officialdom and reached out to Simon. By the time Simon was formally engaged in August by the committee, he had already achieved much in isotope separation and would largely solve the problem by December 1940. Simon, who won the Iron Cross, First Class, for bravery fighting for Germany in the First World War, was four decades later knighted by Queen Elizabeth for services to science, earning perhaps a unique double distinction.

As Simon was progressing his work on the 'gaseous diffusion' of uranium, Peierls had calculated that the correct critical mass (the smallest amount of material needed to sustain a nuclear chain reaction) of uranium-235 was about 8 kilograms if formed in a sphere, or about half that if a 'reflector' was used. As official historian of UK atomic program Margaret Gowing concludes, 'By March 1941 therefore an atomic bomb had in effect ceased to be a matter of scientific speculation'. In Chadwick's words, it 'was not only possible – it was inevitable'.

Peierls' updated calculations and Simon's progress in isotope separation formed the core of the MAUD Committee's interim report and were swiftly shared with the Americans. This vital information might have been expected to provoke the Americans into action. Instead, Briggs who received the memo in March 1941 did not even share it with his Uranium Committee members. He put it in his safe. There it remained – safe but useless.

The same month, researchers using Ernest Lawrence's new 60-inch cyclotron at Berkeley converted uranium-238 into a new element with atomic number 94. The young American chemist Glenn Seaborg and Italian physicist Emilio Segre soon proved that, like uranium-235, it would easily fission. They named the new element 'plutonium', after Earth's most distant planetary neighbour. It too would play a big role in the years ahead – though, again, the discovery was not officially acknowledged for some time.

In July, the MAUD Committee, pulling all available research together, submitted its final report to the British government, stating that the bomb – this new 'super-weapon' – 'was feasible' before the end of the war. But with the Battle of the Atlantic now raging and German U-boats tightening their grip on vital sea lanes of supply, Britain simply had no resources and no capacity. The committee recognised that if the bomb were to be built it – and it had to be built – it must be in the United States.

Some resisted the inevitable. Chadwick dissented from the committee's conclusion, and so did Churchill's scientific muse and adviser, Frederick Lindemann, recently ennobled as Lord Cherwell, who argued that the bomb should be built 'in England or at worst in Canada' so that Britain was not at the 'mercy' of the United States.

Churchill, always an easy mark when Britain's glory was at stake, agreed. 'Although personally I am quite content with the existing explosives', he famously wrote in a memo to his chiefs-of-staff, 'I feel we must not stand in the path of improvement'. Behind the drollery, Churchill's decision to proceed was the first by a national leader to approve the development of an atomic bomb. The chiefs-of-staff soon optimistically agreed with Churchill: sparing neither time nor resources, the bomb should be built in England and perhaps tested on 'some lonely, uninhabited island'. To that end, a new directorate was founded, named 'Tube Alloys'.

It was sheer folly to go it alone. The man in charge of Tube Alloys, Sir John Anderson, realised it almost immediately. He proved to be on the right side of history. If the British program was spending about £430 000 in 1943 on nuclear research and development, the United States was spending nearly 250 times as much. Soon, it would be spending even more. But it took time to convince Churchill and the top brass to accept the inevitable and officially reach out to the United States for atomic cooperation.

While the leaders prevaricated, by mid-1941 most experts, including Thomson and Oliphant, knew the future of the bomb lay on the other side of the Atlantic. In this way, the scientists were more realistic than the politicians. Britain not only lacked sufficient resources, it was fast running out of time to single-handedly take on a project of this magnitude. Bombed and blockaded at home, its armies had been routed in Greece and were being pushed hard in North Africa. In Europe, the Nazi empire now stretched from the Pyrenees to occupied Poland. A month before, three million German soldiers crossed the Russian border in the largest land invasion in history and were smashing the Red Army along the entire front. Britain was no longer standing alone against Hitler, but with the startling success of the German blitzkrieg, for how long?

Yet if the solution and salvation rested with the United States, there was little sign the American cousins understood this cold reality. Neither the Tizard Mission nor Britain's subsequent scientific liaison channelling the work of the MAUD Committee to Washington had so far managed to shake America from its nuclear torpor.

Somebody else, Thomson thought, would need to rattle the cage.

《 • 》

Howard Florey was demanding, and never more than now. His call went out – bring me penicillin and bring it quick.

Elated at the success of the mice experiment, Florey was also intensely frustrated. Now confident that penicillin would work to treat infection in humans, he needed more of the drug for trials to conclusively prove its miraculous healing powers. Much more, in fact, than current techniques, dubbed the 'surface culture fermentation method', could ever hope to produce. The technique, essentially the same as Fleming's, involved growing a mould culture on the surface of a medium of sugar and inorganic salts. After about a week, the active ingredient was extracted from the shallow liquid under the blanket of *Penicillium notatum* and dried as a powder. 'The yield was pathetically low', wrote Heatley many years later, 'in terms of pure penicillin, not more than one milligram per litre'. The task of growing the mould might best have been sub-contracted to a commercial firm to do on an industrial basis, but in wartime that was not possible. The Dunn School would have to do it, and that meant Heatley. He would have to design and build his own factory.

For starters, they needed better vessels. Lab dishes and food tins would no longer cut it. Heatley discovered that the containers most effective for making penicillin were the bedpans from nearby Radcliffe Infirmary, but only a dozen or so were available. So he designed his own modified bedpans and approached a large chemical glassware manufacturing firm with the specifications. Yes, they could make it, but they were prohibitively expensive and production would take at least six months.

But then someone at the lab had the idea of a ceramic vessel. Florey eagerly volunteered a contact with a pottery business and sought their assistance. They could not help but put Heatley in touch with a firm that might. Heatley drove the 200 kilometres north to Burslem, one of the six towns of Stoke-on-Trent in Staffordshire, and was delighted to learn that the potters had made three ceramic prototypes, one almost perfect for the job. Heatley finalised the design with the potters: 'out came the pocket knife', he recalled later. It still looked a lot like a bedpan, but its purpose would be even nobler. Most importantly, it could be cheaply made in large quantities.

Having tested mould samples in the new prototype and found it a great improvement on existing culture containers, a full order was placed. Heatley drove back to Stoke-on-Trent on 22 December 1940 in a borrowed small Ford van used by Chain's wartime blood service and picked up the first 174 of 500 ceramic vessels. He drove back through a snowstorm to Oxford. With four colleagues, he spent Christmas Eve washing and sterilising about half the ceramic pans and then filling each of them with about a litre of mould-growing broth. On Christmas Day, Heatley returned to the Dunn School and with a spray gun seeded the broth with spores of the fungus.

By New Year's Eve, Heatley reflected that the Dunn Lab was now growing 'nearly one thousand times' more penicillin than a year ago. This, in turn, required more hands to help. Florey managed to secure funding from the Medical Research Council, and when Heatley could find no men for the job, he quickly recruited six women to nurture and grow the penicillin mould. They became the first of the 'Penicillin Girls' – young, many only teenagers, poorly paid but enthusiastic and proud of their vital role in upscaling the production of penicillin. Without them Heatley could never have produced sufficient penicillin for human trials.

It was now seven months since the first mice experiment. Even with 500 glorified bedpans, six Penicillin Girls, and some significant technical improvements to the process it still took six more weeks to produce the quantity required for human trials.

Unknown to Florey, Dr Martin Dawson of Columbia University in New York had read with fascination *The Lancet* article and, having obtained a mould culture from the Pennsylvania Department of Agriculture, began injecting patients as early as October 1940. The penicillin doses he could produce, however, were far too small to have any positive effect on his subjects. Florey's was the first credible human trial of the new drug.

Still fearful of the possible side effects of penicillin on humans, Florey decided to test the toxicity of penicillin on a terminally ill patient. Mrs Elva Akers, an Oxford local dying of cancer, proudly volunteered for the experiment. The first injection, by Dr Charles Fletcher, had too many impurities and caused Mrs Akers to run a temperature, but the next injection went well and she had no reaction to the penicillin. Confident that penicillin was not toxic to humans, Florey next sought a very sick patient suffering from an otherwise terminal bacterial infection. He did not have to go far.

《 • 》

Albert Alexander was a 44-year-old constable in the local Oxford county police force and was soon to become one of the most famous patients in medical history. Lying in the septic ward of the nearby Radcliffe Infirmary, where Heatley had only recently scrounged for bedpans, Alexander was dying from a simple scratch to his face (apparently suffered during an air raid and not from the apocryphal rose thorn). The raging infection had already cost him his left eye and was now rapidly spreading from his scalp and face to his lungs and shoulder. He was pus-ridden and covered in abscesses.

Starting on 12 February 1941, Alexander was injected with a course of penicillin. Within just a day, he had improved dramatically and pulled back from the brink of death; within only five, and with the swelling nearly gone, he looked on the road to recovery. Albert Alexander was a miraculous sight.

Then disaster struck. Even with extra penicillin being extracted and recycled daily from Alexander's urine, Florey ran out of doses. With full recovery in sight, Alexander relapsed, dying after a long struggle on 15 March 1941.

Florey and his team were despondent and frustrated. For a few days they had glimpsed the extraordinary healing power of penicillin. But bacteria was a tough opponent that needed a prolonged antibiotic blitz. A teenager treated at the same time as Alexander also showed rapid improvement before the penicillin ran short. Fortunately, despite a renewed infection, the young man survived. A middle-aged husband of a Dunn Lab employee was treated for a carbuncle. While the infection was not life-threatening, Percy Hawkins made a complete recovery. For the first time penicillin had successfully cured a patient of his infection.

Nevertheless, Florey was in a bind: he needed to prove to pharmaceutical companies that his drug worked to cure serious, life-threatening infection, yet without external help he could not produce enough penicillin for trials. Until their methods of production improved, Florey and his team decided to work only on sick children as they needed less penicillin. But even then, it was not smooth sailing.

The last three patients in the trial were two young boys and an infant. The first of these, 4-year-old Johnny Cox, was desperately ill when first administered penicillin. Without it, he would have been dead in a matter of days. Within just three days of starting the treatment, the young boy showed improvement and within nine days was happily playing with toys and penicillin was stopped. But then, tragically and without warning, the boy died. Florey was upset but determined to find out the cause. He did. An autopsy showed that the infection had cleared and the young boy had died of an unrelated aneurism. It had nothing to do with penicillin. Patients five and six, a boy and a baby, both recovered well after treatment.

For Florey and his team the mixed results reflected the limited quantities of the drug at their disposal. They had little doubt that penicillin was the most significant treatment in fighting entrenched infection in humans. Heatley described the results as 'almost miraculous', but they were based only on a few rodent tests and a small clinical trial in six selected subjects, one of whom relapsed after treatment stopped and two of whom died, even if one was from an unrelated cause. A handful of mice, two kids, a carbuncle sufferer, and a teenager who eventually recovered on his own: this was all they had to show for two years of intense work. It was tough going. Today, the process and this record would not pass muster. But these were not normal times – Heatley described this lack of bureaucratic regulation as 'our greatest piece of luck'.

Throughout the first few months of 1941, the Dunn School received a procession of official visitors, chaperoned by penicillin supporters like the Royal Society's president, Sir Henry Dale. All remained uncommitted. It wasn't only scientific scepticism about a treatment that sounded almost too good to be true; the war and the demands it placed on British manufacturing meant there was no spare capacity.

As he had done before when frustrated by a lack of local interest and resources, Florey turned to the Rockefeller Foundation. It seemed like a lifetime ago when the foundation had first found promise with Florey's penicillin project. But he was confident that the results he and his team had achieved since 'Dusty' Miller's visit to Oxford in early November 1939 would justify further support.

Luck was on Florey's side. In mid-April, Warren Weaver, the head of the foundation's Natural Sciences Division was convalescing in London after a car accident. Barely able to contain both his exasperation and excitement, Florey outlined to Weaver his progress so far and the need for more penicillin to trial. Could Weaver and the foundation help?

As a matter of fact, yes, they could. The American immediately grasped penicillin's potential far better than the harried executives of British pharmaceutical firms. Weaver soon told Florey that the foundation would cover his travel expenses to the United States if he could get permission to leave Britain. The Rockefeller Foundation would also arrange meetings with potential partners in the industry, but the rest was in Florey's hands. They would facilitate but not finance. It would have to do.

The lifeline from America came with a steep price. Florey had to choose one colleague to accompany him to the United States. The purpose of the trip was to find a willing manufacturer, and Heatley knew better than anyone the process and challenges of production. So Florey chose Heatley but told no one.

On the morning of departure, Chain saw Florey with bags packed and was quick to question him. Florey, no doubt, had been dreading this confrontation but did not back down on his decision. Chain was left behind.

He never forgave Florey, writing later, 'I left the room silently but shattered by the experience of this underhand trick and act of bad faith, the worst so far in my experience of Florey. It spoiled my initially good relations with this man forever'. Chain always saw penicillin as a joint project between the two of them. All others, including Heatley – a mere technician in his opinion – were unimportant. Professional courtesies between Florey and Chain might remain but the warmth and intimacy of those exciting early days in the hunt for an antibiotic never returned.

8

Missions to America

Cidade da Luz. The City of Light.

Staring goggle-eyed as they made their way up and down Avenida da Liberdade, one of Europe's most beautiful boulevards, the two hot and tweedy scientists were overwhelmed. Having arrived fresh from Oxford via bombed out, boarded up, and blackened out Bristol, Lisbon was bright, white and searing. Heatley described the three days in the Portuguese capital as a 'freely-lit, non-rationed paradise', where locals still indulged with a spoonful or five of sugar in their coffee. While Heatley sampled the exotic fare, Florey worried about the effects of heat on their precious cargo.

They had slipped out of Oxford after lunch on 26 June 1941, carrying with them the seeds of a medical revolution: freeze-dried mould samples, vials of penicillin powder, notebooks full of data and observations, and the typescript of the team's second, forthcoming *Lancet* article. Departing at dawn next morning from the Whitchurch airstrip 'bristling with wire and guards', it took them seven hours and some luck to reach Portugal's friendly shores.

In Lisbon, they were met by representatives of the Rockefeller Foundation, including, Lennard Bickel records, 'a Mr Makinsky'. Aged 40 at the time, the 'suave, dapper' Makinsky was born a khan (a prince) into a noble Iranian family in service to the tsars. After fleeing the Russian revolution as a teenager, he settled in France, and from 1926 worked for the foundation's Division of Medical Sciences in Paris, before moving to its Lisbon office for the duration of war. A brilliant networker who spoke eight languages, he was the perfect contact for the Rockefeller Foundation in Europe. Makinsky gave Florey and Heatley $50 in escudos each, confirmed their flights from Lisbon, via the Azores and Bermuda, to New York for 30 June (in the event, they were delayed a day), and escorted them to their lodgings, the Hotel Tivoli in central Lisbon.

Governed by Antonio de Oliviera Salazar, an academic-turned-autocrat described by one British diplomat as the most physically beautiful of Europe's dictators, Portugal was walking a tightrope of neutrality in a world at war. Rich only in wolfram, a hard metal used to strengthen steel, ball bearings and armour-piercing shells, this small and virtually defenceless country had to balance its historical alliance with England (dating from 1373, the oldest in Europe) with an existential threat from Nazi Germany or Franco's Spain – or, indeed, both. Not an easy task.

Its cafes awash with gossip and bars sweating rumours and intrigue, the Portuguese capital was the initial setting for what became the 1942 cinema classic *Casablanca* and also inspired a host of lesser wartime thrillers like *One Night in Lisbon*. One can only imagine its impact on the two scientists in their heavy English suits and thick ties as they progressed up the palm-lined avenues. Florey could easily have been mistaken for another foreign movie producer scoping the city, as he eagerly recorded on his two cameras (a 35 mm Leica and a 16 mm movie) the bustle of luxury shops and trendy restaurants with their eclectic clientele of locals, refugees, diplomats, spies, spivs and smugglers.

Florey and Heatley were on constant alert in a city that traded in two currencies: escudos and information. Florey had been warned to be careful. He knew Germany was aware of penicillin and, as he and Heatley checked in to their hotel, he scoured the small foyer of the Tivoli for 'idle and solitary men' who might be trouble. He handed his precious briefcase to the manager and stood by until it was securely locked in the hotel safe. He wasn't taking any chances. Though a physician committed to human healing, Florey understood that penicillin would give an advantage to the Nazi war machine. He and Heatley had a whole long weekend in Lisbon, watching over their shoulders and tripping over shadows. Florey checked his briefcase each morning.

Certainly, Lisbon was alive with intrigue. Exactly one year before, the city and its glamorous riviera played host to SS officer Walter Schellenberg, protégé of Reinhard Heydrich, head of the SS counterintelligence branch, plotting to kidnap the Duke and Duchess of Windsor. More recently, Ian Fleming (no relation to Alexander) had left for America on a naval intelligence mission. When he returned to Portugal later in the summer of 1941, he met one of Britain's top foreign spies, the Serbian-born playboy and double agent Dusko Popov, who became his model for James Bond. Meanwhile, Jack Beevor, saboteur, spy and father of the military historian Antony, busied himself preparing a scorched-earth campaign should the Nazis invade Portugal. This was not an unlikely scenario in the summer of 1941, preoccupying *Abwehr* officers at the German military intelligence service's biggest station in Europe. Inside this rat's nest of spies, conspiracies and rumours, Florey was right to feel worried.

It came as a relief when their Pan Am Dixie Clipper seaplane took off from the River Tagus, upstream from the centre of Lisbon, on Tuesday morning, 1 July. Transatlantic flights were still long and precarious, with pilots navigating by stars at night. Yet for all the dangers this was, for civilians, the fastest and most convenient way to get to America. And on a Pan Am clipper, with its dining room and bar and a turn-down bed service, the journey would never, in the history of air travel, be more luxurious. But a well-fed Florey chose to sleep in his armchair instead, the mould samples inside the onboard fridge and the briefcase held tightly on his lap the entire journey.

《 • 》

It was 33 degrees Celsius and humid when Florey and Heatley arrived in New York the following afternoon. Never comfortable with the media, Florey fobbed off a reporter from the *New York Times* who asked him why he had come to America. On 'medical business', he replied, unwilling to say whether it was in an official capacity. Fortunately for the reporter, there were other, more colourful passengers to chase. Miss Catherine Dreyfus, a 21-year-old French actress, claimed that 'her great-uncle, Albert Moissan, was Napoleon's last companion in St. Helena'.

'Medical business' was an understatement. Florey's mission was to persuade the political, bureaucratic and corporate establishments of a neutral country to take a gamble on a drug trialled so far on only a few dozen rodents and half-a-dozen people. It would take all of Florey's genius and advocacy to secure official support given this woefully inadequate case history. Only the Americans could make penicillin happen. The question was whether they would.

Florey and Heatley dropped their luggage at their Madison Avenue hotel and headed straight to the RCA Building in Midtown Manhattan. The Art Deco building housed the headquarters of the Rockefeller Foundation and most other Rockefeller enterprises, including Standard Oil. Warren Weaver, so supportive in London, was back home, fully recovered from his car accident. He scrutinised the scientists' itinerary, hoping that the Rockefeller Foundation's support would be vindicated. With no time to waste, the in-house opportunity to test Florey's penicillin proposal was scheduled for the next day, before the Head of the Medical Sciences Division at the foundation, Dr Alan Gregg.

This was Florey's first chance in the United States to present the miraculous story of penicillin. Tired and under pressure, he performed superbly, confirming his greatness both as scientist and as advocate. Florey was not a practised public speaker, and he was certainly no showman. But as he quietly detailed the Oxford team's struggle to first identify possible biological antibiotic agents and then make a drug to battle infection in humans, Florey's presentation was steeped in drama. No adjectives and superlatives were needed when describing the triumph of the first mice experiments, nor the heartbreak of the revival and then deaths of Albert Alexander and young Johnny Cox.

Heatley, awed by the presentation, years later told Florey's biographer, Lennard Bickel,

> I remember him best of all for that performance ... Even though I knew the subject well, he showed me new facets, and I realized suddenly how great a man he was. None of us in the penicillin team could have matched him, and he was so clearly the leader. I count that hour in Gregg's office as one of the great experiences of my life.

It was gushing testimony from Heatley, but no exaggeration. Gregg was wholly convinced by the display, delighted that the foundation's money had been well spent, and happy to throw his weight behind Florey and the promise of penicillin.

But he warned his guests against unwittingly becoming involved in bureaucratic turf wars or patent battles with local drug companies. Florey accepted the advice, adding that all he wanted in return from American industry was a kilogram of penicillin for his own human trials.

Florey's next American engagement mixed the personal with the professional. With only a few hours notice to their hosts, he and Heatley jumped on the train to New Haven to spend the 4 July Independence Day holiday weekend with Florey's children, who for safe-keeping were staying with his old friend John Fulton, now Professor of Physiology at Yale, and his wife, Lucia. As it happened, Paquita was away at summer camp, but Florey at least reunited with Charles. Fulton was delighted to see Florey and hear of his progress. He instantly got to work teeing up appointments with some of the most influential US scientists and bureaucrats. It says much about Florey's personality and the respect, indeed, the esteem in which he was already held, that so many went out of their way to back him. After all, if not quite snake oil, the new drug was far from proven.

After a long weekend of relaxing and taking stock, Florey and Heatley embarked on a whirlwind of meetings, seminars and soirees. Florey would later compare his experiences in America to a 'man with a carpet bag', going 'from place to place seeking aid'. Selling penicillin rather than plush pile, he sought to duchess, charm and convince a succession of suspicious and secretive senior scientists and bureaucrats to take his miraculous powder seriously. Florey was desperate that they recognise penicillin's potential benefits to humankind, to the United States in the event of war and, not least, to American drug companies themselves. Though Florey would not have thought so, in many ways, this was his finest hour.

Among those cajoled, Dr Charles Thom proved to be the most consequential and immediately useful. Thom was an odd fellow, but – or maybe because – he knew more than anyone about moulds and fungi. A decade before, he had correctly identified Alexander Fleming's original mould as *Penicillium notatum*. Hopeful that Florey and Heatley might indeed discover improved ways to make penicillin in America, he now arranged their visit to the US Department of Agriculture's Northern Regional Research Laboratory (NRRL), an inconspicuous sounding institution in the small Midwestern city of Peoria, about 265 kilometres south-west of Chicago.

It was a lucky break for Florey and Heatley. The NRRL's Director, Dr Orville May, greeted them on 12 July and immediately agreed to pursue ways to increase penicillin yields. Self-assured in the breezy manner of mid-century Americans, May believed that with the expertise of his Fermentation Division, his lab – larger than anything his guests from Britain had ever seen – would soon find a way to mass-produce penicillin.

It took a little longer than May first thought. When Florey left Peoria after five days to continue his trying role as a travelling salesman, the penicillin spores so carefully bought from Oxford were still refusing to germinate. Heatley, who had stayed on to lend his expertise, was on edge. Was Peoria's humid summer to blame, or did the spores die along the way – perhaps as early as in the hotel safe in Lisbon? If the penicillin spores failed to grow and reproduce, he and Florey would be severely embarrassed, and the prospect of mass production gravely set back.

A few days after Florey's departure, Heatley finally spotted a whitish cloud turning blue-green in a culture flask – the first signs of germination. Never had the growth of a humble mould been more welcome. Heatley and his principal collaborator, Dr Andrew Moyer, now began working in earnest. Bristling with native Midwestern isolationism and Anglophobia, Moyer did not make for an easy colleague. Aware of the importance of their partnership for the future of penicillin, Heatley bit his tongue and quietly got on with the job.

Slowly, but with mounting confidence, the fermentation wizards of Peoria found ways to increase penicillin yields. First, Moyer's suggestion they substitute lactose for the sucrose used by the Oxford team boosted the mould's growth. Next they added corn steep liquor, the gluey substance left over after extracting starch from corn. The NRRL, nestled deep in America's Corn Belt, had been tasked with finding use for this abundant agricultural waste. Heatley and Moyer found the liquor, rich in the nitrogen that promoted plant growth, increased the penicillin yield tenfold, well beyond their wildest dreams.

But they needed more. The Oxford team had never managed to overcome the innate inefficiency of using only the surface of a nutrient-rich medium to cultivate the mould. It took hundreds of ceramic flat pans to produce a sufficient quantity of penicillin for a single patient. Heatley did his utmost back in Britain to improve the process, but 'surface fermentation' would never make enough penicillin. Unless they could overcome this challenge all the work so far would be in vain.

《 • 》

While Heatley continued to enjoy cornbread and Midwestern hospitality in Peoria, Florey felt increasingly strained and frustrated on his rounds of drug companies. A naturally reserved man, he did not enjoy schmoozing, charming and cajoling cynical business leaders. 'I have literally met dozens of people + drunk gallons of intoxicating liquors', he wrote to Ethel on 22 July. '[M]y nose is even redder than usual.'

Even with plentiful social lubrication, progress was painfully slow. As Pfizer executive John Smith remarked when first introduced to Florey's penicillin project, 'The mold is as temperamental as an opera singer, the yields are low, the isolation murder, the purification invites disaster'. Drug companies were also wary of investing in cumbersome fermentation technology in case penicillin could be synthesised – made artificially from component chemical parts – more easily and cheaply.

To Florey's growing exasperation, hesitancy was proving the norm. Only a month before, in July 1941, the Office of Scientific Research and Development (OSRD) had been set up to replace the old National Defense Research Committee. The new body had an ambitious mission to expand the scope of research into new areas, including military medicine. At first it looked promising. But as one member of the OSRD's Committee on Medical Research, the biochemist Baird Hastings, recorded after conferring with Florey, 'It was the impression of the Committee that the difficulties of manufacture were so great and the probable cost of treating so large that penicillin could hardly be expected to become a drug of practical significance during this war'.

Florey had one more card to play. For the man put in charge of the Committee on Medical Research was someone that he knew quite well. This lucky coincidence would lead to perhaps the most far-reaching meeting in the history of penicillin.

Alfred Newton Richards was an old acquaintance, dating back to Florey's 1926 Rockefeller Foundation travelling grant to America. The two had got on well and respected each other's intelligence, integrity and judgment. At lunch on 7 August 1941, overlooking the historic Rittenhouse Square in Philadelphia from Richards' social club, Florey recited the penicillin story. Travelling salesman that he had reluctantly become, he was now good at it.

Florey acknowledged that evidence flowing from the human trials was pretty thin, but expressed his certainty that the production problems could be overcome with more money and dedicated research, allowing for more extensive testing in sick patients. He had little doubt these would establish penicillin's revolutionary potential. The data was far from complete, but the story compelling and – most importantly – the storyteller credible.

Richards had once described the younger Florey as 'a rough colonial genius'. Fifteen years later, looking across the dining table at his guest, now in early middle age, hair greying, his care-worn look partly softened by his gold-rimmed glasses, Richards saw someone whose acumen and insight he could trust. If Howard Florey believed it could be done, Richards would back him. Risking his own standing as the committee chairman, Richards would now put his weight – and that of the US Government – behind the project.

It was a big call, the biggest of Richards' life and, after the atomic bomb, the biggest to be authorised by Bush's OSRD. The OSRD's official history recalls that Richards 'encouraged' firms to produce penicillin, even though 'the difficulties which attended its production were so great that nearly two years of work by Florey's team had yielded an amount sufficient to treat only five [sic] patients'. Twenty years later, when Richards was asked why he staked so much on seemingly so little, he replied, 'Florey is a scientist, and a scientist like that doesn't tell a lie'. In the end, it was personal. Richards believed in Florey, so he believed in penicillin.

This act of scientific faith proved momentous. Henry Harris, Florey's successor at the Dunn School, summed it up succinctly: 'Without Richards, Americans would never have taken over production of penicillin'.

History sometimes displays strange symmetry. Just two years before his critical meeting with Richards, Florey had gingerly but hopefully written to Sir Edward Mellanby seeking but a £100 for chemicals on which to grow the Oxford team's first batch of penicillin-producing mould. And just two years after that fateful afternoon in Philadelphia with Richards, enough penicillin was being produced to satisfy the heavy demands of the Allied armed forces locked in war against Germany and Japan. Richards was both midpoint and the fulcrum on which Florey's penicillin project swung from early hope to global saviour.

As Florey left the club that day, he was hopeful but far from certain of Richards' impact. It would, in any case, take months to bear fruit. And so, Florey resumed his melancholy rounds of industry, bureaucracy and academia. Tired and bothered as he was, his firm faith in the magic mould remained intact.

《 • 》

Two days before Florey sat down to lunch with Richards in Philadelphia, Oliphant made his own transatlantic crossing.

What a B-24 Liberator made up in transit time, those onboard paid in sheer discomfort. The B-24 was designed to carry bombs, not passengers. Inside an unheated, uninsulated and unpressurised bomber, conversations were mute and oxygen masks mandatory. Winston Churchill, who frequently flew long distance in a modified Liberator, ordered his oxygen mask to be customised to allow the smoking of cigars, sending jitters through his safety-conscious minders. Unlike the prime minister, Oliphant could not light up onboard. Seated this time on the flight deck rather than down the fuselage, as customary for a passenger, Oliphant had a lot on his mind, and his own comfort and safety were the least of it. The proximity of the crew did not provide much distraction.

The previous summer Hitler, aglow with his victory in the West and hedging his bets on the Luftwaffe's ability to bomb Britain into submission, told his military commanders: 'Britain's hope lies in Russia and America … If Russia is smashed, Britain's last hope will be shattered … Decision: In view of these considerations Russia must be liquidated. Spring, 1941'. Although delayed by a couple of months, the liquidation was now well underway.

As Florey flew to America back in early July, German panzers had taken Minsk. Only ten days into Operation Barbarossa and the Germans were already a third of the way to Moscow. By the time Oliphant climbed onboard the Liberator a month later, the Wehrmacht had broken through Smolensk and was now two-thirds of the way to the Russian capital. The blitzkrieg steamroller might have slowed, but its pace was still awe-inspiring, taking into account the paucity of good roads and the logistical challenges of keeping three million soldiers, 600 000 vehicles and as many horses in the field and fighting. The Americans, more optimistic than other contemporary observers, expected the Soviet Union to resist for another month or three at most. After that, Britain might again expect to be Hitler's target, this time with the resources of the entire continent from the Pyrenees to the Urals marshalled to aid the cross-Channel invasion.

Time was not on Britain's side. In North Africa, the only front where the British Army was in action against Axis forces, fighting continued to see-saw inconclusively. The fortress of Tobruk, with Australian troops inside it, would remain under German siege until November (eventually captured, albeit briefly, by Rommel in June 1942). But if Africa exasperated Churchill, forcing him to fire a succession of underperforming commanders, the far greater and more immediate threat rose up from the depths of the Atlantic. In April 1941 alone, the *Kriegsmarine*'s U-boats and surface vessels sank almost seven hundred thousand tons of British merchant shipping, killing several hundred crew. Such losses were not sustainable, making the prospect of hunger and even starvation very real.

Oliphant's trip to the United States should have been a happy sequel to the Tizard Mission a year earlier, serving to introduce the godfather of the cavity magnetron to his many American admirers. Yet for all the progress in short-wave radar technology achieved on both sides of the Atlantic, a spectre haunted the trip: Germany might already be working on a super-bomb. Britain had the necessary science but no resources; America had the resources but seemingly no interest. Sharing their atomic secrets with the United States had failed to spark collaboration. Oliphant sought to find out why. The very survival of Britain, even the democratic world, might depend on it.

9

Aussie stirrers

Oliphant's arrival in North America got off to a bumpy start when his plane got lost and nearly ran out of fuel, but he was given a hero's welcome. Scientists, bureaucrats and military officials were all delighted to meet the man of 'O. Tube' fame. Oliphant was soon confident that, at least as far as the development of radar was concerned, scientific cooperation between Britain and America was assured.

Coming from wartime Britain, Oliphant was struck by the bright lights and optimism of America's north-east coast. Despite the lingering legacy of the Great Depression and the onset of war in Europe, the Great Republic had found new hope. Like Florey and Heatley, Oliphant welcomed America's rising confidence.

The United States had certainly come a long way in the 11 months since Taffy Bowen and the Tizard Mission had first pulled the cavity magnetron out of their black box. Having thrown talent and money at the challenge, the

Americans were now reaping the rewards. Bell Telephone laboratories became a key centre for developing, testing and manufacturing magnetrons, while MIT's Rad Lab researched better detection techniques. Oliphant stopped by Bell to discuss ways to make magnetrons shockproof, a quality important for their use as instruments for war. He also had another trick up his sleeve. Meeting with the Rad Lab's director, Lee DuBridge, Oliphant plucked out of his case a new 'strapped' magnetron. Developed in his Birmingham lab by James Sayers, it was ten times more efficient than any previous magnetron. The Americans, once again, were awed.

Oliphant journeyed to Washington and the Smithsonian, where radar technology was being used to guide proximity fuzes that exploded shells within a predetermined distance of a target. Another of the Tizard Mission's bounty, the OSRD came to regard proximity fuzes as one of the most important war-winning innovations. Oliphant was proud of the initial British work of invention, and happy to see others forge ahead with improvements and mass production. 'Bugger it all, this is not really interesting work, this is work for engineers now', he later reflected, 'so I went back to work on nuclear physics'.

Oliphant made for an unlikely emissary. True, as an Australian he was a useful outsider, unencumbered by the complicated history of Anglo-American relations, and an early convert to the need for greater scientific cooperation between the two countries. A lively and jovial interlocutor when he wanted to be, he could, however, be easily moved to belligerence if he believed he was dealing with idiocy – which he often did. His abrasiveness and directness in dealings with colleagues and officials made him notorious, but at the same time impossible to take lightly and ignore. Nearly 40 years old, with greying curly hair, ruddy cheeks and round wire-rim glasses, Oliphant retained cherubic and rather innocent looks. But he also possessed the energy of a demon and the forceful personality, the contacts and, on account of radar, the standing to demand an audience. He would not be fobbed off, a crucial quality for a go-between tasked with a difficult mission of utmost importance to the course of war.

While discretion was not part of Oliphant's make-up, he had been asked by George Thomson of the MAUD Committee to make 'discreet inquiries' of his American counterparts and rouse them into action. Oliphant believed the atom bomb could and should be built – and fast. His mission was to convince the Americans to do it.

In perhaps the greatest compliment paid to him by an American, the National Security Advisor to Presidents Kennedy and Johnson, McGeorge Bundy, wrote that Oliphant now 'crossed the United States like a nuclear Paul Revere'. Like the Revolutionary War hero who rode through Massachusetts raising the alarm and alerting his sleeping compatriots, Oliphant dive-bombed into the swamp of American bureaucratic inertia and scientific doubt. The task would try his patience and stretch him to his limits. It would in turn frustrate, bemuse and enrage him; it would also make for his finest hour.

《 • 》

Oliphant's first stop was the Uranium Committee, established two years prior by President Roosevelt. Accompanied by Charles Darwin, he was anxious to see its chairman, Lyman Briggs, but nothing could have adequately prepared him for this meeting. Briggs, a former soil scientist and statistician, was a political appointee nearing retirement. He was tired, sick (he was soon to have a major operation) and looking forward to a quiet life in a big office hidden within Washington DC's bureaucratic maze. Amiable and courteous, Briggs played with his pipe and proudly pointed out a tiny cube of uranium on his desk before detailing the Uranium Committee's progress. Their focus was on nuclear energy as a potential power source, and it looked promising.

But what about the bomb? What have you done about that? Oliphant's mind spun as the bumbling Briggs revealed he had not even looked at the minutes of MAUD Committee meetings, or their reports, including the final one from just a month ago. And neither had anyone else working under him. None of the papers, from the Frisch–Peierls memorandum onwards, regularly passed to Briggs by Fowler and Darwin since the middle of 1940, had been shared. After all, figured Briggs, they were marked 'Most Secret'! It was unbelievable.

Oliphant was 'amazed and distressed' that the chairman of the committee set up to investigate the possibility of a nuclear chain reaction was completely unaware that British scientists had established its viability. Forty years later Oliphant's fury had barely subsided as he wrote of Briggs: 'This inarticulate and unimpressive man had put the reports in his safe and had not shown them to members of his Committee'. It was an unforgivable dereliction of duty. 'Here it was September 1941, nearly three years since we knew about fission', Oliphant reflected, 'Britain and the Commonwealth fighting for life, depending on America to avert the horror of Hitler being first with the atomic bomb, and all that precious time lost'. The Uranium Committee, established following Einstein's intervention in 1939 and tasked by Roosevelt to 'study blowing people up', had had two years to make a bang. Instead, it barely made a whimper.

《 • 》

Oliphant now proceeded to do what Briggs should have done: brief the Uranium Committee members on the ongoing work of British scientists. The time for professional courtesies and diplomatic niceties had passed.

He made quite an impression. His friend, John Cockcroft, reported that the term 'Oliphantic' had already been coined in Britain to describe Oliphant's outspoken behaviour. Discretion be damned, Oliphant now turned his blunt Australian manner on the unsuspecting Americans.

Samuel Allison of the University of Chicago recalled Oliphant peering at them through his glasses:

> he said 'bomb' in no uncertain terms. He told us we must concentrate every effort on the bomb and said we had no right to work on power plants or anything but the bomb. The bomb would cost twenty-five million dollars, he said, and Britain didn't have the money or the manpower, so it was up to us.

The committee was stunned, believing its task was to develop nuclear power for submarines.

Oliphant was appalled, and he said so. Bad enough that Briggs had kept the committee in the dark. But here was the United States' government body charged specifically with looking into the military application of nuclear fission and not one of its members had seriously contemplated the possibility of nuclear weapons. He would have surely agreed with one future historian who concluded that 'since being created in 1939, the Uranium Committee saw its feeble, dithering, unsavoury reputation continually deteriorate'. Leo Szilard argued later that the Uranium Committee's inertia and misplaced priorities had set the atom bomb project back by at least 18 months. Bearing in mind the last 12 months of the war proved the deadliest for German and Japanese soldiers and civilians, deploying an atom bomb to force unconditional surrenders in Europe and Asia in mid-1944 might have saved millions of lives. No one could have foreseen that in August 1941, but the potentially revolutionary impact of an atomic bomb should have been obvious to any intelligent observer – as it had been to Frisch and Peierls when they wrote their memorandum, and subsequently to other scientists and policymakers in Britain.

Yet it was not. After the Briggs and the Uranium Committee fiascos, Oliphant decided to redouble his efforts, convinced that the Americans, even when informed about atomic weapons, did not fully understand the feasibility and urgency of the task.

Next on Oliphant's appointment list was James Conant, Briggs' former boss at the National Defense Research Committee (and the NDRC's new chair, after it was subsumed under the OSRD in June 1941). Conant was a brilliant chemist who gave up research and the likely prospect of a Nobel prize for the Presidency of Harvard University, in the process becoming perhaps the most influential university administrator in American history after Woodrow Wilson. A tall and slim bespectacled Boston native, Conant was a self-confessed Anglophile and regarded by the British and Oliphant as a friend at the Roosevelt court.

It should have been a happy and fruitful meeting. Conant later recalled that by the summer of 1941 he was 'impatient' with physicists on the Uranium Committee. Although initially sceptical, he claimed he had already been convinced (by the Harvard physical chemist George Kistiakowsky) that an atom bomb could be developed in time to play a part in the current war. But, he wrote, he kept it to himself. His conversion to the cause was cemented mid-year with 'the news that a group of physicists in England had concluded that the construction of a bomb made out of uranium 235 was entirely feasible'.

The talk of 'news' is strange. There was nothing new about it. Like Briggs, Conant and his boss, Vannevar Bush, had been regularly supplied with copies of MAUD Committee documents by British diplomats since mid-1940. Conant later claimed that 'the first I had heard about even the remote possibility of a bomb' was from Lord Cherwell while visiting London in March of 1941. That very same month the MAUD Committee had concluded in its interim report that an atomic bomb was 'practicable and likely to lead to decisive results in the war'. Apparently, neither Conant nor Bush were reading up on the continuing progress of MAUD scientists either. Was anyone? Oliphant might well have wondered why Britain even bothered to pass on their findings.

If, in the summer of 1941, Conant was finally a bomb believer, he showed no signs of it in his meeting with Oliphant. The Australian's direct and forceful advocacy for immediate action to build the bomb fell on polite but unsympathetic ears. The tight-lipped Conant would not even engage with his visitor on the current state of American fission research. Recalling in 1985 his meeting with Conant, as well as a subsequent one with Bush, Oliphant concluded, 'they were not interested. They said, oh that's for the next war not for this one'.

In a postwar memoir, Conant did not mention Oliphant by name. But even if given a cold shoulder in public storytelling, Oliphant made a large impact. In his 1943 secret paper, *History of the Development of an Atomic Bomb*, Conant wrote that the 'most important' reason for the sudden change in direction of American atomic policy in the autumn of 1941 was the fact that 'the all-out advocates of a head-on attack on the uranium problem' – Oliphant named first and foremost among them – 'had become more vocal and determined'. Still, Conant did nothing after first meeting Oliphant.

Even a fellow physicist, the legendary Italian-American Nobel laureate Enrico Fermi, was unmoved by Oliphant's pleading. In a twist of fate, just two years later Fermi was to be responsible for conducting the world's first controlled nuclear chain reaction, a critical step to the success of the Manhattan Project. But in September 1941, he could not be persuaded, Oliphant reporting that 'Professor Fermi was non-committal about the fast neutron bomb'. Technical problems seemed too great and there was no certainty, argued Fermi, that fission would cause an explosion.

The only encouragement that Oliphant received on the American East Coast was from the eminent physicist and Director of General Electric's research laboratory, William Coolidge. Coolidge listened intently to the animated Oliphant. He was stunned. Just a couple of months before, he had reviewed Briggs' Uranium Committee and endorsed its progress and findings – or lack thereof. That conclusion now seemed premature. But Coolidge, like the committee members, had not seen the succession of MAUD Committee reports stacked neatly and unread in Briggs' safe. Moved by Oliphant's revelations of British progress, Coolidge wrote to Frank Jewett, the President of the National Academy of Sciences, suggesting that 'Oliphant's story should be given serious consideration'. Finally a receptive ear. But it was not enough; Oliphant had to grab the attention of those with real power: America's top science administrator and his political master.

《 • 》

Sitting atop the American scientific bureaucracy was Vannevar Bush, director of the newly created OSRD and effectively the country's chief wartime scientist. Bush had ultimate oversight of the Uranium Committee. More importantly, he had the ear of President Roosevelt. Oliphant, well aware of Bush's importance to his mission, arranged to meet him in New York. He took some confidence from Bush's appreciation of technical scientific problems. Bush had been President of the Carnegie Institution for Science, and before that Dean of Engineering at MIT. He was no bureaucratic layman.

Oliphant described the rapid development of nuclear science in Britain, pausing to emphasise the groundbreaking work of Frisch and Peierls. He admitted that he had been a sceptic before their work convinced him. He also pointed out that the MAUD Committee had largely confirmed their findings. An atomic bomb had ceased to be a matter of scientific speculation. The question was no longer 'if' but 'when' – and, more worryingly, 'who' and 'where'.

With almost regal rectitude, Bush did not admit to any knowledge of the MAUD Committee's work. He was aware Oliphant had led the team that refined microwave radar for military purposes – probably the most significant new technology fine-tuned and mass-produced under the OSRD's aegis – but the pair had never met before. Bush was far more cautious than Oliphant about discussing secret nuclear research. Oliphant queried whether there was any research to be secretive about.

There were later claims that even prior to meeting with Oliphant, Bush was concerned about the Uranium Committee's lack of progress and would soon push for an independent assessment – the third one in a year – of America's nuclear research program. One American historian even goes so far to 'guess … that Bush was secretly delighted' by Oliphant's lobbying, as he sought to secure support to persuade Roosevelt to build the bomb. If so, Bush did not share this with Oliphant – or anyone else.

Oliphant thought the much-anticipated encounter was a failure. He was not able to convey to Bush the sense of urgency that now propelled the British nuclear research effort. He could hardly be bothered to remind Bush that if an atomic bomb were to be built it would have to be in North America. After 20 frustrating minutes, Bush terminated the meeting.

« • »

Further north, Anglo-American relations were warming up. Aboard the HMS *Prince of Wales*, Churchill steamed in mid-Atlantic towards the coast of Newfoundland to meet with President Roosevelt aboard the USS *Augusta*. For Churchill the meeting was a triumph. While not securing America's entry into war, he left with guarantees of massive aid to assist Britain's war effort. Roosevelt, in return, sought assurances about the postwar world. Their agreement was struck on eight important principles of freedom, self-determination and economic cooperation. After they were made public on 14 August, these principles became known as the Atlantic Charter, and eventually formed the basis of the United Nations Charter in 1945. After more than a year of 'crushing disappointments and reverses', recalled his bodyguard Inspector Walter Thompson, Churchill seemed to allow his troubles to 'sink into the deep ocean we had traversed'. Roosevelt and Churchill hit it off. For their subordinates it was not always so easy.

Florey arrived back in Peoria on 15 August 1941 to check on Heatley's progress. After the initial hiccups and false starts, they had promising news. A few days before, Moyer and his colleagues showed Heatley how to grow mould using the 'deep-tank fermentation' method. The principle, familiar from brewing, was simple enough: a culture was placed in large vats and continuously injected with air and stirred to promote the growth throughout the whole submerged mixture. If successful, this would produce a quantum leap in quantity from Heatley's old 'surface fermentation' techniques.

Submerged in tanks, the mould spores now developed overnight. But what about the quality? It would take some time to test the potency of the deep fermentation–produced penicillin. Judgment was still out.

The day after Florey's return to Peoria, the results of his human trials at Oxford were finally published in *The Lancet*, in an article titled 'Further Observations on Penicillin'. A non-descript title once again belied a medical revolution, but there was no disguising the favourable implications of the trials detailed inside: 'Enough evidence, we consider, has now been assembled to show that penicillin is a new and effective type of chemotherapeutic agent, and possesses some properties unknown in any antibacterial substance hitherto described'.

It would take time for copies of *The Lancet* to cross the Atlantic, but Florey had been free and fast with the draft of his article, sharing it widely on his American rounds. He had used it as a ready aid for his evangelism.

From Peoria, Florey proceeded to Toronto in Canada to visit the highly regarded Connaught Laboratories. With Canada already at war with Germany, and Dr Ronald Hare, Fleming's old penicillin collaborator at St Mary's, now working at Connaught, Florey was expecting a positive reception. But his hopes were dashed by the laboratory's director, Dr Robert Defries. Like American business executives, Defries feared production through fermentation was too difficult and the chemical synthesis of penicillin likely at any time. He was also concerned that, despite initially promising results, penicillin might work only against staphylococcus or at best against a very limited range of bacteria. Florey thought Defries bumptious in rejecting collaboration and from then on maintained a dislike of the Canadian.

Some researchers, however, were happier to receive Florey. The pioneering Dr Martin Dawson, who a year earlier was the first in the world to administer penicillin to a human subject, hosted Florey in New York. Florey recognised a familiar but melancholy sight – hundreds of bottles brewing mould in classroom laboratories at Columbia University. Been there, done that; it's not enough.

After two months of knocking on doors, Florey was fed up and glad to be heading home. On 10 September, he and Heatley met for a debrief with Warren Weaver in New York. Dizzy with people and places, Florey duly reported on the trip and the use of Rockefeller money. It had been a hell of a ride since July, but in truth he was disappointed. He had received no commitments that penicillin would be produced. Sure, some in the pharmaceutical industry had expressed hope that the time would come soon, as had a few in government. There were germs of plans for something to happen sometime in the future. But who knew when? Certainly not Florey.

Heatley's frustrations were of a more personal nature. While committed to staying in Peoria until 16 December, he was finding work with Moyer increasingly difficult. Heatley noticed Moyer was hiding results from him. The two had agreed to co-author a paper on their work, and when a few months later Heatley passed his draft to his colleague, Moyer assured Heatley he would get back to him with his suggestions. He never did.

Time for a breakthrough of any sort – either technological or political – was fast running out. As the investigation into fermenting penicillin in Peoria wore on, patience was wearing out. Despite significant early gains at the NRRL, administrators were starting to grumble. Given their other priorities, was penicillin worth the gamble? Even as late as 29 September 1941, the assistant chief of the USDA's Bureau of Agricultural Chemistry and Engineering, HT Herrick, wrote to Charles Thom that there was 'too much planned for the Fermentation Division to give any further consideration to the production of Penicillin'. The dawning of the age of antibiotics, like that of the nuclear age, wavered precariously.

« • »

After more than six weeks running around the United States' eastern seaboard, charming and cajoling all who would meet with him, Oliphant too was in despair. He knew his mission to get America to consider the implications of the MAUD report, let alone confront the inevitable reality of an atomic bomb, had so far failed. People like Bush and Conant were the apex of the new breed of American science administrators elevated by the coming of the Second World War. But with the United States still languidly at peace, bureaucratic inertia and buck-passing were smothering Britain's desperation and stifling Oliphant's initiative.

Oliphant had no permission, and therefore no funding, to travel to America's west coast. But he knew he had to go. Like Florey he faced bureaucratic torpor and ignorance and he too had a final card to play. The catalyst for the trans-continental journey was the one man in America Oliphant knew would help, his old friend Ernest Lawrence – Lawrence the 'livewire'.

Oliphant had first met Lawrence in 1933 at Cambridge and spent time with him in America in 1937 and 1939, comparing notes about the building of a cyclotron, an 'atom smasher', which Lawrence invented, and the discoveries that such technology would allow. 'They were as alike as two peas in a pod', said physicist James Tuck. 'Oliphant knew a little more physics than Lawrence, but both were energetic developers, essentially promoters'. A keen and well-connected American promoter was just what Oliphant needed.

Tall, broad-shouldered with slicked-back strawberry blond hair set behind rimless glasses, Lawrence was a self-confident and assertive scientific impresario. He could not 'tolerate laziness or indifference' in those who worked with him and, as Oliphant recalled, possessed a presence that was 'at once noticed' and 'profound'. He was impossible to ignore. 'The boy wonder of American science', he had recently won the Nobel prize and, importantly, was now engaged in fission research (plutonium, which his team had just discovered, would become the fuel for the second atomic bomb, dropped on Nagasaki). His self-confidence, entrepreneurship and drive, combined with his charm, contacts and unparalleled access to America's scientific elite, set him apart from the rest of the pack. He was no diffident and reserved boffin. Together with Oliphant's blunt and forceful powers of persuasion they made a formidable team.

Lawrence, who rarely took holidays, was in southern California with his family when Oliphant arrived in San Francisco, but quickly returned to Berkeley to see his friend. On 23 September, he drove his Australian guest along the twisting dirt road to the summit of Charter Hill, among the green hills above the Berkeley campus, where the magnet for his giant new cyclotron was being built. They paused to look out across a beautiful view of a hazy San Francisco Bay. The clearing was surrounded by eucalyptus trees and Oliphant breathed in the distinctive fragrance of his distant homeland. He then turned to Lawrence in bubbling frustration.

Oliphant recounted his meetings with Briggs, Conant, Bush and Fermi. No assurance as to the technical feasibility of the bomb seemed sufficient to move the science mandarins, he told Lawrence. And no warning, no matter how dire, of the possibility of a Nazi bomb would spur them into action. Lawrence was disappointed but not surprised. He had encountered similar difficulties in stirring the Uranium Committee to work. But as he listened now to Oliphant, he became increasingly worried. With no access to the MAUD Committee's reports, he had never previously heard of the Frisch–Peierls memorandum, now 18 months old. There was a cool breeze off the bay on this bright early autumn day, but Lawrence was sweating in anticipation as Oliphant revealed the inner secrets of the MAUD Committee's work.

It was a major security breach. Lawrence did not have the required security clearance and Oliphant was not authorised to divulge nuclear secrets to unauthorised individuals. But without Oliphant's desperate and necessary indiscretion, nuclear history would have taken a different course. Almost certainly, an atomic bomb would not have been built in time to end the war in August 1945.

Oliphant outlined the progress of research in Britain and his own conversion from sceptic to believer following the work of his colleagues Frisch and Peierls. The MAUD Committee was now sure that the bomb would be built. It was inevitable. Lawrence, not without some reservation, agreed. He pointed out that his own research tended to support the MAUD Committee's findings. If nothing else, the United States should now independently assess the plausibility of a nuclear weapon. And it needed to do so quickly. While not yet fully convinced of the science, Lawrence was already committed to the cause.

Back in Lawrence's Berkeley office, the conversation continued – frustration at American officialdom, anxiety about what this might mean for the course of war, and optimism that the United States, if fully onboard, could build the bomb before Germany. As Oliphant and Lawrence discussed the implications of the MAUD Committee's report, they were joined by a colleague of Lawrence's,

Robert Oppenheimer. Thin, dark, and charismatic though volatile, the pipe-smoking Oppenheimer had studied under JJ Thomson (the father of George) at the Cavendish, where in a moment of emotional turbulence he tried to poison Oliphant's future friend and Nobel laureate, Patrick (later Lord) Blackett, with a chemical-laced apple. Oppenheimer had departed before Oliphant's arrival in 1927 so the two had never met. He was now stunned by the frankness of the exchange between Oliphant and Lawrence about a classified topic. Years later, the 'Father of the Atomic Bomb', as he would be dubbed by the American press, recalled his unexpected introduction to the subject of nuclear weapons. It was 'an indiscretion – an eminent English [sic] visitor started talking to Lawrence and me … And clearly, his source of his confidence was the work of Peierls in England. And he said it was terrible that Fermi and I were not involved in this' in any official capacity. It was all news to a shocked Oppenheimer, but consistent with Oliphant's view that this was properly a matter for discussion among Allied physicists, government secrecy be damned.

Oliphant had ventured to Berkeley to convince Lawrence of the science of the atomic bomb so he could advocate for it among America's scientific and bureaucratic elite. He succeeded. He also unwittingly bought Robert Oppenheimer to the atom bomb project. Oliphant was the first to make him aware of the contemporary science and politics of the bomb. From then on Oppenheimer could never escape the bomb – neither the fame nor the burden. Oppenheimer might have been confused about his interlocutor's nationality and startled by his candour, but the seed of destiny was sown.

Compounding his initial indiscretion, Oliphant wrote to Oppenheimer soon after their meeting, summarising their discussion. And going beyond science, Oliphant now pitched the politics: 'Whichever nation is first to succeed in this quest will undoubtedly be master of the world. If peace were to come tomorrow it would still be necessary to obtain the answer first, at all costs, for in the hands of a resentful or unscrupulous nation such power would be dangerous'.

Oliphant – again breaching security – also provided a useful summary of the MAUD report for Lawrence's use. Flushed with Oliphant's revelations, the 'boy wonder' immediately phoned the pioneer of gamma ray research and influential scientist, Arthur Compton, at the University of Chicago. He informed Compton that certain developments had convinced him that it might be possible to build an atomic bomb and warned that it was likely the Germans were pursuing the same objective. Stung into action, Compton promised he would organise a meeting at his home with Lawrence and Conant at the end of September. Oliphant was elated. The Americans were starting to listen.

10

The arsenal and pharmacy of democracy

Florey arrived back in Britain on 6 October 1941. He was spent, frustrated and troubled – and still unsure whether he had made any difference. Four days earlier, the Wehrmacht launched Operation Typhoon, expecting it to be its final offensive to take Moscow. The fate of war and penicillin both hung in the balance.

But no one, Florey included, had counted on the energy, determination and sheer heft of Alfred Richards. After his lunch with Florey in Philadelphia in early August, he had become America's great penicillin champion and benefactor, patiently laying the groundwork within his own administrative domain, and then broader bureaucratic and commercial interests. Richards' Committee for Medical Research soon resolved to progress Florey's penicillin proposal and scheduled a general meeting for 8 October. Richards invited key scientific, bureaucratic and commercial players, including representatives of the four drug companies that had shown interest in penicillin – Merck, Pfizer, Squibb and Lederle.

Vannevar Bush, as OSRD head, chaired the meeting. Roosevelt's science chief spent the war years managing and overseeing some 2500 projects that harnessed science to the cause of Allied victory – his mantra: 'Will it help to win a war; *this* war?' The three most consequential involved the development of revolutionary breakthroughs made in Britain by teams led by Florey and Oliphant: penicillin, microwave radar (and radar spin-offs, including the proximity fuze) and the atom bomb. If Bush ever noted this curious coincidence, which was apparent by early October 1941, he left no record of how he felt about these two intense and insistent Australians bearing gifts.

The meeting opened inauspiciously. The government's main 'mould man', Charles Thom, was not effusive in his support of penicillin, detailing its instability, the difficulty of achieving meaningful yields, and the most perplexing problem of all – scaling up production. Unless a means were devised to mass-produce penicillin, still a big 'if' at this stage, it would remain a boutique remedy – a medical novelty – not available in general hospitals, let alone on a battlefield.

Thom's tribulations did not blunt Bush and Richards' message to the gathered pharma bosses. The science generals were clear. Whatever the technical difficulties and unknowns, the penicillin project now had the government's blessing. The drug companies' work towards producing industrial quantities of the drug was in the national interest and so any collaboration between the four (and with the NRRL in Peoria) would not offend America's tough anti-trust laws. Get to it, Bush and Richards urged. America needs you and it needs penicillin. For all their initial reluctance and scepticism, penicillin would prove to be a watershed and a boon for the industry, turning middling drug companies into the giants of big pharma we know today.

It had taken Florey and his team nearly two years to produce barely enough penicillin to treat six patients, two of whom had died – a wafer-thin case load then, and in modern terms non-existent – but Florey's reputation within the scientific establishment and his friends in high places were just enough to convince those who mattered. 'Fortunately', the official OSRD history concluded, 'Richards appreciated the potential importance of the drug and the vigor and imagination with which he promoted its production were regarded as entitling him to the greatest credit in making penicillin available for use during the war'.

All the misery of life as a travelling salesman, with its endless drinks, dinners and despondency, was not in vain. After months of dogged advocacy, worry and frustration, Florey had finally gotten his wish: America was committing itself to penicillin.

《 • 》

Oliphant too could return to Britain having successfully engaged in an Australian specialty. He had been a 'stirrer'. He had perhaps pricked the conscience of Bush and Conant, and sparked the interest of Coolidge, but, most importantly, he had lit the fire within Lawrence. Oliphant could not yet be sure of success. It would take time before he would learn whether he had managed to persuade the Americans to build an atomic weapon. But he had done all he could. He was back home in mid-October.

In the end, it was a close-run thing. As he promised Lawrence, Compton organised the meeting in Chicago with Conant a few days later, at the end of September 1941. It started badly. 'Conant was reluctant', Compton recalled years later. 'As a result of the reports so far received, he had concluded that the time had come to drop the support of nuclear research as a subject for wartime study'. Conant was adamant that research efforts should be restricted to where they could bring results during the course of the war.

Passionate and bubbling with frustration, Lawrence now took the floor. He repeated what he had learned from Oliphant. There was no longer any doubt. The bomb could now be built. That's what the British believed and, having studied their conclusions, he now believed that too. Lawrence then echoed Oliphant's fears that the Germans were racing towards a bomb that would decide the war, while the Americans were still languidly debating the science of it in a manner of a graduate seminar rather than as a matter of civilisation's very survival. Inertia, lack of technical understanding and perhaps, most of all, an inability to appreciate the urgency of the task, had so far led to a failure to act. But not for much longer. 'Conant', writes Michael Hiltzik, Pulitzer Prize–winning journalist and author, 'had come to Chicago still sharing Bush's scepticism about an atomic weapon … [but now] changed his mind'.

For Compton, the meeting was 'the start of the wartime atomic race'. He credits it as the moment 'We in the United States … saw for the first time that exploration of the possibility of atomic bombs was a military necessity for the safety of the nation'. Lawrence was a powerful and persuasive advocate and Conant left the meeting a convert, determined in turn to convince his superiors (most importantly, Bush) of the importance and the urgency of the nuclear cause. For his persuasive and timely advocacy as the United States stood on the verge of cancelling the uranium project outright, Hiltzik dubs Lawrence 'the man who saved the Manhattan Project'. Yet it was indisputably Oliphant who first moved Lawrence. Oliphant had found the right man for the job.

Now that there was will, there was a way. And a pace – it only took another week or so for the wheels to move after the Chicago meeting. On 9 October, coincidentally just a day after tasking drug companies with mass production of penicillin for the war effort, Bush was briefing President Roosevelt and Vice President Henry Wallace at the White House about the implications of the MAUD Committee's findings. Just to be sure, these were re-transmitted yet again in time for the meeting by Thomson himself, fresh from his role as the committee's chair and now British scientific liaison officer in Ottawa. Bush informed Roosevelt of the conclusions of British scientists that an atomic bomb with an explosive force of some 18 000 tons of TNT might be ready as early as the end of 1943. But the cost would be high. Massive industrial effort would be required to construct a plant to produce enough fissionable uranium-235 and even then, success could not be guaranteed. Bush, however, was no longer a sceptic. Conant, invoking Lawrence and Compton, had persuaded him that the United States must quickly determine for itself whether the British were right – whether a bomb could in fact be built. Roosevelt agreed.

Back in April 1941, Bush asked Compton and the National Academy of Sciences to appoint two committees: one of physicists and the other of engineers to review uranium research independently of the Briggs' Uranium Committee. Both committees focused mainly on nuclear power – boilers rather than bombs – and were unable to make recommendations about the possibility of an atomic bomb. With no weapon in sight, some committee members insisted that 'the entire uranium project should be put in wraps for the duration'. Now, following the White House briefing, Bush asked Compton and the academy to have yet another look.

The academy convened on 21 October. Lawrence commenced by reading Oliphant's summary of the MAUD Committee's final report. Also present at the meeting was

Robert Oppenheimer. He had heard it all before – straight from the horse's mouth – and was fully onboard with the call to atomic arms. The third academy report was delivered in November and this time its conclusions largely accorded with the views of Oliphant and the MAUD Committee. The American establishment was finally onboard, all singing from the same Frisch and Peierls hymn book – a bomb could be built. The president would soon decide whether it would be built.

On 6 December, working at his Berkeley laboratory, Lawrence succeeded in separating one millionth of a gram of uranium-235 from uranium-238 using his cyclotron. On the same day, and now with explicit presidential imprimatur, S1, as the Uranium Committee had been renamed, officially grasped the nettle of the atom-bomb program in the United States. And not a moment too soon. The next day, at 7:48 am local time in Hawaii, and 1:18 pm in Washington, Japanese bombers bypassed snoozing American air defences and unleashed their explosives and torpedoes on the ships of the Pacific Fleet anchored at Pearl Harbor. For nearly four years, from the day of 'infamy' until Japan surrendered, the United States would bring its military and industrial might to the cause of war. Science would be at the apex of this might.

America's entry into the war also guaranteed penicillin's future. Richards held his follow-up conference with pharma companies just ten days after the Japanese attack. It was all go. The same sense of urgency that had fired Florey and Oliphant to sell their scientific wares on behalf of besieged

Britain was now shared by the United States. War galvanised American officials to relentlessly pursue the bomb and penicillin, as they had already done with microwave radar a year earlier. Stirred from slumber, the industrial and creative might of the United States would now embark on a crash course to turn the scientific visions of the two Australians into a war-making and war-winning reality.

The Americans would also try to take the full credit, nowhere more so than in the creation of the atom bomb.

《 • 》

In the early 1960s, in a cornflake box in the strong-room of Britain's Atomic Energy Authority, researchers found a three-page typed foolscap document entitled 'On the Construction of a "Super-bomb"; based on a Nuclear Chain Reaction in Uranium'. A little later, former English war correspondent and writer Ronald Clark located another three pages entitled 'Memorandum on the Properties of a Radioactive "Super-bomb"'. For 20 years, these six pages that together make up the Frisch–Peierls memorandum lay forgotten in classified obscurity in government archives, Peierls' typed-up foolscap gathering dust – and cereal crumbs. Despite being familiar to key British and American players in the nuclear genesis, the document was written out of the early official histories of the nuclear project.

Instead, chroniclers of the bomb's origins lay a straight path between Einstein's letter to President Roosevelt of August 1939 and the president's decision to build an atomic bomb in late 1941. Developing the atomic bomb became an all-American enterprise, made possible only by the brilliant scientific, engineering and industrial effort of the 'arsenal of democracy', as Roosevelt dubbed America's new global role. Lieutenant General Leslie Groves, the man in charge of the Manhattan Project, credited British desperation and a sense of urgency but not their scientific contribution. 'I cannot escape the feeling that without active and continuing British interest there would probably have been no atomic bomb to drop on Hiroshima', he wrote in his 1962 memoirs, but described the British contribution as merely 'helpful'. Oliphant found Groves personally likeable, but whether or not Groves liked Oliphant, he did not find him consequential enough to warrant a mention in his memoir of the development of the atomic bomb.

But General Groves is positively gracious in comparison to the anonymous author of the United States' Official History of the Manhattan Project completed just after the war. Not only were British scientists described as of 'moderate attainments' but their contribution was judged as

> in no sense vital and actually not even important. To evaluate it quantitatively at one percent of the total would be to overestimate it. The technical and engineering contribution was practically nil. Certainly it is true that without any contribution at all from the British, the date of our final success need not have been delayed by a single day.

This is ludicrous. Without British science the Manhattan Project would have commenced later and cost more Allied (and probably Axis) lives. That, at least, is certain. American statesman McGeorge Bundy was happy to credit the MAUD report and its impact on Bush and Conant as the 'immediate catalyst' of Roosevelt's decision to proceed with the bomb. 'British science', concluded Bundy, 'had won admiration in the Washington of 1941. Roosevelt knew the story of radar, and he had warmly supported both Bush and Conant in their early commitment to closer collaboration with British war science'. Interviewed in 1980 the Nobel laureate and physicist Isidor Rabi had no doubt that it was the commanding prestige of British science, their scientific judgment and counsel that was the 'controlling element' in the American decision of late 1941 to build the bomb. And if it was the Frisch–Peierls memorandum confirmed by the MAUD Committee that provided the immediate catalyst for America's decision to build the bomb, it was Mark Oliphant who had made it impossible for the Americans to ignore that science any longer.

The truth of the matter, Oliphant wrote to Sir Edward Appleton in October 1941, was that 'I, personally, was responsible for the whole of the recrudescence of this subject and had to fight hard to get things going and to get Frisch and Peierls accepted'. When their memorandum was finally published in 1964 as part of the official history of the British atomic research program, the celebrated journalist Chapman Pincher wrote that the question 'Who really invented the atom bomb?' could be finally answered. Later, a British official historian described the memorandum as the most 'fascinating and impressive' document in nuclear history.

It remains also perhaps the most awesome. A half-dozen foolscap pages written in austere English laid bare a secret of nature with momentous implications. In first demonstrating the feasibility and practicability of nuclear weapons, global politics, international relations and armed conflict were forever altered. We continue to live every day with the consequences of Frisch and Peierls' calculations and revelations.

The MAUD Committee was in turn specifically set up to test – today we would say 'peer review' – the memorandum's initial conclusions. The committee's assessment that the atomic bomb was 'feasible' first galvanised the British and subsequently helped to bring the Americans onboard. Without the memorandum, no MAUD; without MAUD, no Manhattan Project or, at least, not then. There is little doubt that the Americans would have eventually – months or even years later – decided to build the bomb, but it would have come too late for the millions of young American servicemen required to take the Japanese home islands by conventional means.

The story of the American bomb is not a straightforward path from Einstein to Roosevelt; it is a scientific, administrative and military maze. Einstein's letter did not lead to the desert sands of Los Alamos but to the bureaucratic swamps of Washington DC. The Uranium Committee led nowhere for two years, until Oliphant stirred the pot. The Einstein letter described vague possibilities, while the Frisch–Peierls memorandum and the MAUD Committee confirmed that a bomb could be built to win the war.

'One good way to understand the force and originality of the Frisch–Peierls memorandum', wrote McGeorge Bundy, 'is to consider what might have happened if Frisch and Peierls had composed it in Germany and delivered it to a German government'. Both scientists were terrified of such a prospect – recall Frisch's fear that if Denmark, his temporary refuge, was overrun by the Germans, he would be 'compelled in some way or other to work for Hitler'. Thanks to Oliphant's foresight, Frisch and Peierls wrote their startling conclusions in English.

From rescuing Frisch from Hitler's grasp and encouraging Peierls to Birmingham, through supplying them with resources against official opposition, championing their memorandum to Tizard and then the MAUD Committee on which he served, all the way to, perhaps most crucially of all, badgering the Americans to understand the inevitability of an atomic bomb, Mark Oliphant was the central and indispensable figure. The one constant from the theoretical genesis of the bomb in March 1940 until his return from the United States in October 1941, on the eve of the American commitment to building the bomb, Oliphant pushed mightily for Anglo-American cooperation, realising before most that the bomb could be built, that it had to be built, and that only America could do it.

Postwar histories, whether of an official or personal nature, largely ignored Oliphant's role. Britain's official 'Statement Relating to the Atomic Bomb' published in 1945 mentions Oliphant's role in Tube Alloys and his later work with Lawrence in isotope separation. Unsurprisingly, perhaps, it makes no reference to his critical advocacy both in Britain and the United States in 1940 and 1941.

Recognition took decades to emerge. One of the catalysts was the publication, in 1964, of the official history of British atomic research. It singled out Oliphant's advocacy in America and his 'decisive' influence – essential in energising Lawrence's great talents. The view of Oliphant as the man who started the chain reaction has become the historical consensus. In his 1987 Pulitzer-winning definitive history, *The Making of the Atomic Bomb*, Richard Rhodes concludes (with echoes of Henry Harris' views of Florey's contribution to penicillin), 'Oliphant convinced Lawrence, Lawrence convinced Compton, Kistiakowsky convinced Conant. Conant says Compton's and Lawrence's attitudes "counted heavily with Bush"'.

But even early on, some keen observers, perhaps less invested in questions of national pride and scientific precedence, understood Oliphant's crucial role in the nuclear saga and were not hesitant to give credit where it was due. The principal author of Einstein's August 1939 letter to Roosevelt, Leo Szilard, was one of them. After the war, he recalled how:

Oliphant came over from England and attended a meeting of the Uranium Committee which neither Fermi nor I was permitted to attend ... He realized something was very wrong and that the work on uranium was not being pushed in an effective way. He travelled across this continent from the Atlantic to the Pacific and disregarding international etiquette told all those who were willing to listen what he thought of us. Considerations other than those of military security prevent me from revealing the exact expressions which he used.

'If Congress knew the true history of the atomic energy project', Szilard concluded, 'I have no doubt but that it would create a special medal to be given to meddling foreigners for distinguished services, and Dr Oliphant would be the first to receive one'.

One of the last survivors of the Manhattan Project was the distinguished physical chemist Professor Lawrence Bartell. Recruited at 21 years of age to the atomic-bomb program, he was interviewed in 2013, the still sprightly 90-year-old clearly recalling both the false dawn of the US atom-bomb program and its subsequent kick-start:

> What [Roosevelt] did [in late 1939] was to ask the Head of the National Bureau of Standards [Lyman Briggs] to form a committee to get going. The Head of Bureau of Standards was first of all sort of ill at that time, and

second of all, sort of a wimp. So nothing much happened. It was only after Frisch and Peierls figured out that it was feasible to get a bomb that [Britain] sent a distinguished Australian physicist named [Mark] Oliphant to America to wake up America.

'He woke up America very effectively', concluded Bartell, 'and the rest is history'.

11

The 'salvation of the Allied cause'

Winston Churchill couldn't sleep.

He would later confess, 'The only thing that ever really frightened me during the war was the U-boat peril ... I was even more anxious [about this battle] than I had been about ... the Battle of Britain'. He knew that if German submarines were successful in severing the lifeline to North America, the dominions and colonies, Britain would likely lose the war. For Germany, the Battle of the Atlantic was their best – and, as it turned out, their last – hope of bringing Britain to its knees.

The Third Reich was prevailing in the Battle of the Atlantic as late as March 1943, successfully strangling supplies for Britain, hampering Allied aid to Russia, and preventing the assembly of troops and materiel necessary for the opening of the Anglo-American Second Front in France. Yet by May, the Atlantic had become Hitler's watery

Stalingrad, and Allied air forces began to bring the war home to German industry and civilians.

At the heart of this astonishing strategic turnaround was the 'pearl beyond price' – the cavity magnetron. Oliphant's breakthrough technology gave pilots the ability to see the enemy clearly for the first time, day or night, whether it be a surfaced submarine or its conning tower emerging from the waves, an aircraft to be intercepted, or a ground target to be bombed. It also enabled artillery, particularly anti-aircraft batteries, to direct their fire more accurately. In air, at sea or on ground, the importance of microwave radar for the course of the Second World War cannot be overstated.

《 • 》

From July to October 1940, as British fighter pilots were crippling the Luftwaffe's offensive capability (if at high cost, with 1977 German aircraft destroyed to the RAF's 1744 in the Battle of Britain), German submarines sank 274 Royal Navy and Merchant Marine vessels – a million and a half tons of shipping – for the loss of only six U-boats, just two of them in combat. The commander of the German Navy's U-boat Arm, Vice Admiral Karl Doenitz, boasted that his U-boats could win the war by themselves. Perhaps unsurprisingly, Doenitz's men referred to this period as the 'Happy Time'.

British losses were devastating. Each 7000-ton ship sunk was the equivalent of a 'mile-long freight train falling into the seas, taking with it enough supplies to feed, clothe and fuel a small city or an army division for three weeks', not to mention a crew of 50 to 80. Air cover for British shipping was meagre and frustratingly ineffectual, as RAF pilots vainly scoured the seas trying to spot U-boats. This was difficult during the day, and at night or in bad weather virtually impossible. The aircrafts' limited range out of the airfields of North America and Britain also left large stretches of the mid-Atlantic – dubbed 'The Greenland Gap' or 'The Black Pit' – completely bereft of Allied air presence and protection.

The prospect of Britain running out of ships well before Germany ran out of submarines haunted the Admiralty throughout 1941. After the United States joined the war against Germany in December that year, the United States Navy and Merchant Marine threw their significant numbers behind Britain. It provided some relief but brought no end to the slaughter on the high seas. If anything, the sinking of the *Norness* by U-123 within sight of the bright lights of New York Harbour in the early morning of 14 January 1942 marked a fresh wave of destruction. For four months up and down the eastern seaboard, German U-boats terrorised American shipping. Doenitz named it 'Operation Drumbeat' but for his cocky commanders it was the 'Second Happy Time' or, more menacingly, 'American Shooting Season'. Allied shipping losses in 1942 dwarfed those of the previous two years.

It took the United States months to introduce even the most basic safety measures like blacking out coastal cities so that ships did not stand out as clear targets against a lit-up shore. By July 1942, with U-boat kills now starting to dwindle along America's east coast and escorted convoys increasing across the North Atlantic, Doenitz withdrew his U-boats to 'The Black Pit'. The orgy of destruction they left in their wake in the course of Operation Drumbeat claimed 5000 lives and 400 merchant vessels representing half a million tons of shipping, sunk at the cost of just 22 German submarines.

《 • 》

By the time of the Tizard Mission in September 1940, Britain had developed nearly all the key components of microwave radar. In America, Taffy Bowen played a similar role to Norman Heatley a year later, outlining the course of British research, explaining the technology, and pointing to possible ways of improving it. The acclaimed physicist Isidor Rabi recalled:

> Everything we knew at the beginning about radar we learned from [the British] ... They were tremendous, both technically and what I call philosophically. How to use it. What do you need and why do you need it and how would you use it when you got it? What are its limitations? What are its strengths? ... The British could do wonders with a few sets.

But it was now America's job to take microwave radar from a miraculous prototype to a mass product fit for war, and they threw themselves headlong into this project.

The 'salvation of the Allied cause'

The Rad Lab at MIT was at the heart of this national enterprise. It opened in late 1940, under the inspired leadership of Lee DuBridge who, with support from Ernest Lawrence, hastily recruited from among America's top nuclear physicists. Karl Compton, president of MIT and older brother of Arthur Compton, called the Rad Lab 'the greatest cooperative research establishment in the history of the world'. DuBridge certainly had an eye for talent, with five of his team subsequently winning the Nobel prize. For now, he set three immediate goals: to develop microwave radars for airborne interception by fighter planes, for long-range navigation by bombers, and for more accurate direction of anti-aircraft guns. By the spring of 1941, across a network of institutions that also included Bell Laboratories, the Naval Research Laboratories and the Army Signals Corp, successful models had been developed for all three purposes and handed to industry to manufacture.

With mass-produced microwave radar sets rolling off the assembly lines of Western Electric (the production branch of AT&T) and Raytheon in the second half of 1941, they finally began to enter service in sufficient numbers to make a difference. Microwave radar first made its mark where it was originally most wanted – in the battle-torn skies over Britain. Before long, one commander of a German night-fighter squadron was complaining to his superiors that the British fighters were picking off his men at will 'like currants in a cake'. The Luftwaffe no longer ruled the darkness.

The 'O. Tube' continued to evolve, spawning many ingenious and successful offshoots. As Allied bombing of German cities became more intensive, microwave technology was applied to improve bombing accuracy, especially at long range. In late 1941, Britain's Telecommunications Research Establishment used the cavity magnetron to develop a revolutionary ground-mapping radar codenamed H2S. When fitted to RAF heavy bombers such as the Stirlings and Halifaxes, H2S 'Blind Bombing Radar' allowed the identification and bombing of targets at night and in all weather. Though initially the War Cabinet considered the cavity magnetron inside H2S so secret and valuable, it did not allow airplanes equipped with this radar to fly over enemy territory. After outcry from RAF pilots, this policy was soon abandoned. Radar beacons were also built to assist bombers safely back to base even in the worst conditions.

Radar was also the vital component of the proximity fuze: a tiny radio transmitter fitted into the nose of a projectile – most often an anti-aircraft round – set to trigger it close to a target. It sounded crazy, but the amazing feat of miniaturisation was achieved and proved deadly effective. The basic design arrived in the Tizard Mission's black box with the cavity magnetron. After further development by the Americans, 22 million proximity fuzes were produced during the war, with the cost per unit eventually dropping to a meagre $18. Vannevar Bush concluded that 'if it had not been for radar and the A-bomb, this would have been hailed as the greatest technical accomplishment of the war'.

The 'salvation of the Allied cause'

But perhaps the most significant difference on the battlefield came when microwave radar was pitted against submarines.

《 • 》

By early 1942, U-boat crews started to wonder why their world had become a little less secure. With three submarines all of a sudden lost in unexplained circumstances, Doenitz later recalled 'a fear that the British had developed unknown to us some anti-submarine device'. He was right, though at first he had no idea what it was.

Through 1941, the Royal Navy began equipping its surface vessels with High Frequency Direction Finding Equipment. HF/DF – or Huff-Duff, as it became colloquially known – was first developed by Robert Watson-Watt in the mid-1920s and used to detect lightning. Now reconfigured, it allowed a bearing to be taken from a U-boat's radio signal so an Allied ship could be sent in pursuit. With similar limitations to the long-wave Chain Home system, the submarine hunter still had to visually locate the target – a feat difficult during the day, and all but impossible at night. The addition of microwave radar onboard ships and, even more importantly, onboard aircraft was the game changer. From 1942 on, every pilot possessed an extra set of eyes that could 'see' a submarine's conning tower at long distance and in darkness, effectively quadrupling the area 'visible' in flight. Radar gave aircraft the air-to-surface ability to spot the needle in a haystack. The bombers used it mercilessly.

The Germans did not take long to discover it was Britain's 'pre-eminence in microwave radar' that accounted for their turn in fortune. 'In areas where air cover was strong', wrote Doenitz, 'our most successful method of waging war would no longer be practicable'. His response was to pull the U-boats away from the coasts and towards mid-ocean, out of range of the land-based aircraft.

There, U-boats waited for the Allied convoy to appear on the horizon. At first sighting, Doenitz would order the nearest submarines to converge. The feared 'wolf pack' shadowed the convoy by day, striking it under cover of night. In the mid-Atlantic, the dark, low and sleek U-boat hulls were virtually invisible on the surface. They used guns and torpedoes against the convoys neatly silhouetted against the night horizon. The Black Pit now became the wolf pack's lair and a graveyard of British and American merchant navies.

《 • 》

In 1942, U-boats sank more tons of shipping than the Allies managed to build and 1943 looked set to eclipse it as a record year. The North Atlantic was the only theatre of war where Germany was still on the offensive, Doenitz's U-boats the last best hope for victory. On 31 January 1943, the German Sixth Army had surrendered at Stalingrad, and on 9 March, Rommel left North Africa in defeat after a last failed counterattack on Montgomery's Eighth Army. It took another two months to finish mopping up the remnants of the once mighty Afrika Corp and their

The 'salvation of the Allied cause'

less feared Italian allies, but the British and the Americans (who had joined the fight in Africa in November 1942) had control of much of the Mediterranean and were readying to invade Sicily. German cities, meanwhile, were prey to more accurate and deadly Allied bombers equipped with microwave radar, launching from Britain and, later, from captured Italian airfields.

The Third Reich had to win the Battle of the Atlantic and came closest 'in the first twenty days of March 1943', according to the British Admiralty's assessment. One hundred and seven merchant ships went down at a cost of only 11 U-boats. It was, British historian AJP Taylor argued, 'the worst month of the war for the Allies in the Atlantic'.

But, in the course of a single month, Germany lost the battle on the high seas. U-boat crews called it 'Black May'. Very long-range Liberator bombers, equipped with larger fuel tanks and armed with microwave radar, started to patrol the area previously beyond the reach of land-based anti-submarine aircraft. The Black Pit was finally filled in; the wolf pack had nowhere to hide. The 'effect', said Air Marshal Sir Jack Slessor, Air Officer Commanding Coastal Command, was 'instantaneous and dramatic'. To add to German misery, Allied naval ships, including active aircraft carriers, were also standard-fitted with microwave radar.

Huff Duff, the breaking of Germany's Enigma code, and a more efficient coordination of aircraft and warships all played their role in the defeat of the U-boat threat. But, as the official British naval historian Captain SW Roskill concluded, 'the centimetric radar set stands out above all other achievements because it enabled us to attack at night and in poor visibility'. His American counterpart concurred.

So did Doenitz. By the end of May 1943, he knew it was all over. Half of the 56 U-boats lost in April and May had been destroyed by radar-equipped land or carrier-based aircraft. Such losses could not be sustained. Doenitz acknowledged, 'Radar, and particularly radar location by aircraft, had to all practical purposes robbed the U-boats of their power to fight on the surface. Wolfpack operations against convoys in the North Atlantic … were no longer possible'. He concluded, 'We had lost the Battle of the Atlantic'.

Uncharacteristically introspective, Hitler agreed, conceding that radar was the only weapon that stood between him and victory in the Battle of the Atlantic. The small device that first lit a car headlight in a university backroom now lit up an entire ocean and sent predators scurrying back to port. U-boats would never again wreak havoc on Allied shipping.

《 • 》

It was an ironic outcome for the nation that was an early leader in radar innovation. After all, German scientists first hypothesised the use of electromagnetic waves for detecting objects and were the first to try it in practice. By the 1930s, they had even developed a magnetron, but discarded it for insufficient transmitting power. At the beginning of the war, Germany had superior long-wave radar technology to Britain but a lack of coordination rendered it much less effective. This defensive technology did not sit right with what radar historian Robert Buderi called Germany's 'military mindset': the obsession by Hitler and his commanders with lightning attack.

Corporate rivalry as well as licensing and patent issues further slowed down the adoption of new radar technology, while excessive secrecy inhibited the German war machine from drawing on the excellent research skills of its universities. In contrast with a desperate British government, which conscripted scientists into service, German radar work was carried out in siloed secrecy. Research institutions, industry and each of the armed services jealously guarded their turf from each other.

But despite institutional and cultural barriers, German radar technology still sporadically shone with brilliance. While British long-wave radar detected and kept at bay German bombers during the day, the Luftwaffe developed increasingly sophisticated radio navigation to more accurately bomb British cities at night. As the carnage of these night 'Blitz' raids increased, Britain responded with their own radio signals jamming or deceiving the Germans. This to and fro was dubbed by Churchill the 'Battle of the Beams'.

Germany never did catch up to the Allies, lagging in radar technology until the end of the war. Reichsmarschall Goering, upon learning of the simplicity and ingenuity of the cavity magnetron captured from a downed Allied bomber in early 1943, conceded:

> We must frankly admit that in this sphere the British and Americans are far ahead of us. I expected them to be advanced, but frankly I never thought that they would get so far ahead. I did hope that even if we were behind we could at least be in the same race.

As the Second World War broke out, Japan was not far behind in radar technology either, inventing a low-powered cavity magnetron before Randall and Boot. The Japanese failed to build on their early success for two main reasons. Firstly, because of inter-service rivalry between the army and the navy, the technology was not effectively shared. So deep did this mistrust run that Japan's main radar plant had to operate two totally separate production lines: one for army radar sets, the other for the navy.

Secondly, and not unlike the German strategic doctrine, the Imperial Navy was hell-bent on attack and saw radar as a defensive technology. It was best suited for installation on the Japanese home islands, but not a top priority for the modern samurai warriors striking in the air or at sea. Most Japanese naval ships were not fitted with radar until late 1943, and even then it was relatively primitive. The Germans copied the captured Allied cavity magnetron in 1943 – though they had no industrial capacity to produce it in numbers that would make a difference for the Luftwaffe

– and gifted the technology to the Japanese. It made its way onto Japanese fighters in the spring of 1945, by which time there were hardly any fighters left in the imperial arsenal.

With the 'Germany first' strategy adopted by the Allies, the Pacific theatre was not the priority for American radar-makers. Initially, the less sophisticated metre-long radar alerted the US Navy to impending air attacks at the Battle of the Coral Sea and then at Midway. The Rad Lab started to consider the specific needs of the Pacific war only in late 1943. Around the same time, Taffy Bowen arrived in Australia to help the government's Radiophysics Laboratory at Sydney University build hardy but lightweight radar sets for use in the Pacific Islands. In early 1944, the Rad Lab had opened a branch office in Australia, sending expertise and support to speed up deployment of microwave radar in the final offensive against Japan.

After a slow start, the full weight of radar technology made a powerful impact. Japanese convoys became easy targets for America's Liberators equipped with 3-centimetre radar. Working in concert with submarines, the low-altitude bombers played havoc with Japanese supply lines. So successful were American airplanes with their state-of-the-art microwave radar that between July and October 1944 they sank 250 000 tons of enemy shipping off China and Vietnam alone, with the loss of only five aircraft.

The critical role of radar in the Pacific was perhaps best illustrated in the Battle of the Philippine Sea, the largest aircraft carrier battle in history, fought on 19 and 20 June 1944.

When the United States launched an amphibious invasion of the Mariana Islands, US naval radar detected scores of Japanese fighters coming their way. American fighter aircraft and then naval guns firing proximity fuze–fitted ammunition shot down 40 of the 69 fighters of the first wave before they came within sight of the convoy. Wave after wave of Japanese aircraft attacked the American Fifth Fleet, and wave after wave were destroyed in what became known as the 'Great Marianas Turkey Shoot'. Of more than 700 Japanese planes at the start of the battle, four in five were lost in combat. A Japanese admiral later credited radar as the single greatest American strength. Japanese aircraft were as good as American ones and Japanese pilots as skilled and courageous, but they could not overcome the enormous advantage that radar – and the proximity fuze – gave their American enemy.

By late 1944 the United States had established air bases on Guam and the Mariana Islands of Saipan and Tinian. From there, the US Army Air Forces' B-29 Superfortress could fly the round trip to Japan. Previously, the USAAF had concentrated on precision bombing, but with the appointment of the stout, cigar-chomping Curtis LeMay, precision was no longer required. Known to his pilots as 'Iron Ass', LeMay quickly realised that largely wood-and-paper Japanese cities were highly susceptible to fire. So, he decided to burn Japan to the ground.

On the night of 9 March 1945, guided by air-to-surface radar, 334 aircraft dropped half a million incendiaries on central Tokyo. It was the most destructive air raid in history, eclipsed not even by dropping of the atomic bomb five months later. More than forty square kilometres of Tokyo, a quarter of the capital, were vaporised in a tsunami of fire, flame and ash. As the bombing continued, thousands ran into the streets 'almost mad with terror'. Mothers fled with their babies on their back, some oblivious that their child's hair was on fire. Others were torn from their parents' grasp by the Akakaze, or red wind. Hardest hit were the docklands where nearly all the wood and paper homes of working-class inhabitants burned to a cinder and lucky survivors jumped into the Sumida River. All that remained 'were iron safes standing forlorn amid the ashes of the homes and offices to which they belonged'. In a few short but deadly hours the inferno killed more than a hundred thousand Japanese, mainly civilians, and rendered over a million homeless.

By mid-June, LeMay had run short of incendiaries, the largest Japanese cities burned to ash. With little opposition – few enemy fighters still able to take off and ineffective anti-aircraft artillery, their radar guidance jammed by the Americans – the United States owned the skies over the home islands. Armed with onboard microwave radar and soon a secret new super-weapon, the B-29s were now poised for ever more terrifying missions.

《 • 》

Of all the technological innovations of the Second World War, the cavity magnetron–powered microwave radar had the greatest bearing on the course of warfare. As radar was radically transformed into a compact and versatile weapon of attack, nearly a million units came into use in the air, at sea and on land. It ensured Britain and America won the Battle of the Atlantic and allowed Britain to become freedom's 'unsinkable carrier', from which the Allies could fight, roll-back and eventually defeat Germany, and later Japan.

This is why the resonant cavity magnetron developed by Oliphant's team and mass-produced for war in the mighty forge of American industry is widely regarded as the Allies' most crucial tool of war. The English physicist, CP Snow, claimed the cavity magnetron was 'the most valuable English scientific innovation in the Hitler war'. For Taffy Bowen, who first dazzled the Americans with it, the magnetron was the 'salvation of the Allied cause'. The Americans did not demur. James Baxter, the official OSRD historian, remarked in his Pulitzer Prize–winning book, *Scientists Against Time*:

> This revolutionary discovery, which we owe to a group of British physicists headed by Professor M. L. Oliphant of Birmingham ... [was] the most valuable cargo ever bought to our shores. It sparked the whole development of microwave radar and constituted the most important item in reverse Lease-Lend.

For Alfred Loomis, to whom Bowen had so spectacularly demonstrated the workings of the magnetron, it was simply 'the breakthrough'. Norman Fine, an electronics engineer and historian, puts it even more bluntly: 'the resonant cavity magnetron was, hands down, the single hardware invention that was most influential in winning the war'.

The events of mid-1943 signalled these postwar conclusions. With U-boats defeated and the lifeline to the world secure, it was now near certain that Britain would hold on and, together with the United States, eventually launch the Second Front to liberate Europe from the west. 'If the British victory at the Battle of El-Alamein in November 1942 had marked "the end of the beginning"', writes historian and broadcaster Jonathan Dimbleby, 'then the German defeat in the Battle of the Atlantic in May 1943 was most certainly "the beginning of the end" – and it was of decidedly greater significance'.

An early enthusiast of radar, Churchill understood its importance to victory. 'All through the war', he reflected, the Battle of the Atlantic 'was the dominating factor'. For three long years it was the U-boats that, more than anything else, kept him up at night. And then, in a space of just one month, the deadly stealth of Doenitz's wolf packs was vanquished and the mass carnage they wrought was no more. The road to victory lay open.

Churchill slept soundly again.

12

Of mice and men and melons

Howard Florey might have convinced Alfred Richards, and Richards, in turn, might have convinced American drug companies of the promise of penicillin, but it was not enough. Towards the end of 1941, Fleming's original mould, so carefully nursed across the Atlantic by Florey and Heatley, continued to frustrate the efforts of scientists seeking to coax a better yield. Every 2000 litres of culture fluid produced only enough penicillin to treat a single case of blood poisoning, or as American doctors told General Eisenhower's PR aide, Lieutenant Commander Harry Butcher, in mid-1942, it 'Takes eleven acres of mold to cure the scorched face of one flier'. This was no way to make a drug for use on the battlefield, where hundreds of thousands of doses would be needed.

Despite different techniques, all of which had increased penicillin production, it was still nowhere near enough. It was too laborious and inefficient to cultivate Fleming's *Penicillium notatum* on the surface of bedpans, and it was now apparent the strain was too capricious and feeble to grow well in deep tanks.

Fortunately, Peoria's finest were not convinced that the mould Fleming stumbled upon was the only choice. There were many other strains of *Penicillium* mould out there (even if only one out of the thousand in the NRRL's own collection had similar properties to *notatum*). Perhaps a different one would be blessed with a more plentiful and potent active ingredient. So the call went out for mould samples to be bought to the eager scientists at the NRRL. The United States Army Transport Command obliged this 'mould rush' and retrieved samples of strains from across the globe, including China, India, South Africa and even Australia.

Some did prove better than others. Yet the answer to the hopes and prayers of penicillin hunters was to be found in the last place they looked: just around the corner from the laboratory.

Salvation came from a young woman named Mary Hunt, a lab assistant at the NRRL. She was tasked with scouring local fields and markets for rotting fruit and vegetables, which earned her the perhaps inevitable nickname 'Mouldy Mary'. One Monday morning in 1943, Hunt arrived at the labs with a musty cantaloupe, or rockmelon, covered in a rich golden mould. It was another flash of serendipity in the penicillin story.

Hunt had struck gold. A mould called *Penicillium chrysogenum* immediately produced 200 times more penicillin than *notatum*. After zapping it with X-rays and later ultraviolet light, the resulting mutant strain yielded an astounding 1000 times the amount of Fleming's original penicillin mould. Even better, *chrysogenum* grew well in deep-fermentation tanks. Production increased rapidly. While only a 1000 people had been treated with penicillin in early 1943, by D-day in mid-1944 there was enough to treat every soldier in need. By 1945, a million people had received the life-saving new drug. Mary Hunt's mouldy melon, arguably the most significant piece of fruit in history, became the source of the penicillin on the beaches of Normandy and much of the rest of the world for years to come (even if, as in many other aspects of the penicillin saga, Hunt's role is contested, with one of the NRRL scientists later claiming the cantaloupe in question was donated by an anonymous Peoria housewife).

Florey's promising but temperamental penicillin, with its wafer-thin case history of success, could now be produced at scale. The Oxford team made the scientific breakthrough, Peoria market provided the melon, and American scientists working in collaboration with the pharmaceutical industry found a way to do what many thought impossible: in a little more than two years they turned an experimental laboratory treatment into a life-saving drug for the masses. Ultimately, 36 universities and hospitals, 22 private companies, 300 labs and 6000 scientists made up the research and production machine assembled by the OSRD and the War Production Board to tackle the penicillin challenge.

《 • 》

By late 1941, Norman Heatley was assisting with penicillin production at Merck and was waiting for his collaborator, Andrew Moyer, to send his corrections to a draft joint paper on their work in Peoria. The corrections never came. Moyer submitted the paper in his own name taking sole credit for work he had done in collaboration with Heatley. But worse was to come. Moyer also applied for and won British and other international patents on production methods developed in Peoria. So, as Ernst Chain had sadly predicted from the outset, the British had to pay royalties on penicillin made using techniques developed at the instigation of Florey and Heatley and with the latter's input and participation.

But in truth Florey did not have much choice. He went cap in hand to the United States in the summer of 1941 because no British firm was willing or able to commit to manufacturing penicillin. The patents covered new techniques developed in America to make penicillin more efficiently and on a mass scale. There was not much of commercial value that might have been patented from the original laborious process Heatley developed at Oxford. British medical ethics also precluded Florey from applying earlier for a patent.

To a few detractors none of this mattered. Their story was that Florey naively gave away penicillin to the Americans who then sold it back to the British for a handsome profit. But in not pursuing the tussle around patents and, later, publicity, Florey's standing in the scientific community continued to soar.

While Heatley stayed behind in America for nearly a year, Florey arrived back in Oxford on 6 October 1941 to pick up where he had left off: trying to produce more penicillin at the Dunn School for another human trial. Although he received assistance from two pharmaceutical companies, the Dunn School would remain the largest source of penicillin in Britain for another two years. The kilogram of penicillin he had asked for in the United States in exchange for his troubles never arrived.

Ethel Florey was put in charge of the second clinical trial, involving 187 cases of sepsis. She was responsible for selecting patients, measuring treatment, organising tests and keeping notes. It was hectic and she was on call for weeks at a time. With her children overseas and her husband rarely at home, she threw herself into her work. She loved it. The clinical results were a triumph that exceeded even her husband's best expectations. He was at first critical that some of the patients chosen seemed beyond healing and might skew the results. But the new drug cured even the most hopeless cases. 'My wife is doing the clinical work', Florey wrote in August 1942 to his old mentor, Sir Charles Sherrington, 'and is getting astonishing results – almost miraculous some of them'. Penicillin had become, in his words 'a corpse-reviver'. Ethel Florey's treatment of patients and collection of relevant data was lauded by all.

In March 1942, Florey was elected a Fellow of the Royal Society (though not principally for his work on penicillin) and a year later, as Doenitz and his U-boats were enjoying their last hurrah in the Battle of the Atlantic, published in *The Lancet* of 27 March 1943 the triumphant results of the trials. Florey planted his flag on the summit: penicillin had been proven; he had arrived. Or so he thought. But glory – and the credit – was not yet his.

« • »

If asked who 'discovered' penicillin, most people will answer Alexander Fleming. Fleming is nearly universally credited with the scientific discovery of the century. Fleming discovered the antibacterial properties of a strain of mould, *Penicillium notatum,* but, unable to isolate the active ingredient, moved on to other substances, including sulpha drugs. He did not discover how to make the life-saving drug: penicillin as properly understood.

Science is a competitive profession and priority for discoveries or inventions a much sought – and fought over – honour. Often precedence is not clear. Randall and Boot did not invent the cavity magnetron, but they did produce the first model powerful enough to make microwave radar a reality. Frisch and Peierls did not invent the atomic bomb, but they were the first to clearly and conclusively show that an airborne atomic bomb was feasible. And Florey and his team did not discover that *Penicillium notatum* was a naturally occurring antibiotic mould, but they succeeded in making a powerful new drug out of it.

Moulds had been known for thousands of years as a folk remedy for infection. Their antibacterial properties had been commented on by a host of eminent scientists, including Louis Pasteur, Joseph Lister, John Tyndall – and finally Alexander Fleming. Fleming progressed science by noticing the antibiotic effects of the penicillin mould in a Petri dish, making a crude mould juice to wash utensils and treat superficial infection and, critically, recording and publishing his observations.

But there he left it. Why? Because he realised, the task ahead was monumental and the risks enormous. 'The forest of uncertainty' stopped Fleming, as it had so many others before him, from pursuing the promise of penicillin the mould to penicillin the drug. 'Notwithstanding claims made years later with the benefit of hindsight, including those of Fleming', wrote Florey's wartime colleague, Dr Trevor Williams, 'it seems clear that by 1935 there was not one person in the world who believed in penicillin as a practical aid to medicine'.

All that changed in 1939, seven years after Fleming had abandoned his work on the mould. Howard Florey and his team picked up the challenge and, after a great deal of labour and luck, made the drug that saved millions. It was, in the words of Andrew Hargadon, 'a 2000-year history of an idea followed by a two-year history of turning that idea into a reality'.

The reason Fleming is the father of penicillin in the popular mind lies in public relations.

When Florey wrote his buoyant letter to Sherrington on 2 August 1942, reporting that 'there is, for me, no doubt that we have a most potent weapon against all common sepsis', penicillin was still virtually unknown to anyone outside a small circle of scientists in Britain and the United States. But within a few weeks it would become a *cause célèbre*. For just as Florey was enthusing about penicillin to Sherrington, 52-year-old Harry Lambert lay dying in St Mary's hospital of a meningococcal infection. Now in a coma, at best he had a few days to live. But he was lucky. He was a friend of Alexander Fleming.

As a last desperate measure, early on Wednesday morning, 5 August 1942, Fleming called Florey in Oxford urgently seeking penicillin for his friend. Florey obliged, immediately catching a train to London with all the penicillin he had available. The only condition Florey placed on Fleming was that he and Ethel could include the results as part of their human trials. Fleming readily agreed and Harry Lambert became case No. 12 of 187. Like all others, he lived. But in his miraculous recovery from a certain death lay the seeds of misplaced credit, abiding antipathy and subsequent myth.

Someone at St Mary's hospital leaked the good news to the London *Times*, which, in an editorial on 27 August 1942, referred to penicillin's 'alluring prospects' and 'strong antibacterial powers'. Echoing earlier calls from *The Lancet*, the paper called for government support of industry's production of penicillin. Other newspapers picked up the story over the weekend, but none mentioned the name of scientists involved.

Then, on Monday, 31 August, in a letter to *The Times*, the lasting legend was born:

> Sir,
> In the leading article on penicillin in your issue yesterday you refrained from putting the laurel wreath for this discovery round anyone's brow. I would, with your permission, supplement your article by pointing out that ... it should be decreed to Professor Alexander Fleming of this laboratory. For he is the discoverer of penicillin and was the author also of the original suggestion that this substance might prove to have important applications in science.

The letter was signed by Fleming's old boss, Sir Almroth Wright.

Almost immediately, Fleming and St Mary's hospital were besieged by reporters, jubilant that amid the bleak news of the war in Africa, Russia and the Atlantic, there was finally a cause for celebration – the possibility of a life-saving medication to balance the daily butcher's bill of death and destruction.

Florey was appalled. He knew that public demand for the 'miracle drug' could not yet be satisfied. Despite outstanding initial results, penicillin was still subject to clinical trials, which were consuming all the available supply. But that no longer mattered. Lord Moran, Churchill's private physician and a former dean at St Mary's Medical School, touted Fleming's 'discovery' for the money it would attract to the hospital, while his friend Lord Beaverbrook, the press baron, made sure Fleming was accorded the 'laurel wreath'.

The truth is, as Florey's close colleagues and supporters like Edward Abraham acknowledged, Florey's 'limited skills in diplomacy' did not help his cause. 'In many ways', his son Charles later reflected, 'it was as much my father's fault that he was denied credit as Fleming taking advantage of the situation'. Florey treated the press with disdain. When reporters turned up one day at the Dunn School, Florey told his secretary to send them away, saying he might be able to give them ten minutes next week. He never pursued recognition for himself or his team through the media, believing that the conventions of British science precluded such tawdry self-promotion. And so he never forgave Fleming, aided and abetted by St Mary's and its boosters, for what he regarded as unwarranted, misleading and even dishonest claims concerning the development of penicillin. It was an 'unscrupulous campaign' to appropriate the scientific credit, he wrote in an angry letter to Mellanby in June 1944, 'with the upshot that [Fleming] is being put over as the "discoverer of penicillin" (which is true) with the implication that he did all the work leading to the discovery of its chemotherapeutic properties (which is not true)'. It did not help that both the mould and the drug made out of it were called penicillin. Perhaps if Florey and his team had chosen another name for the drug they invented, debate about appropriate credit would never have occurred.

Fleming loved the attention and the adulation, ultimately riding the antibiotic revolution to world fame. His failure to give proper credit to Florey and the Oxford team led to an irrevocable breach between the two. But the silver lining was that Fleming and Fleet Street, for all their inaccuracies and myth-making, focused the government's attention on penicillin.

On 25 September 1942, roughly three years after Mellanby promised Florey £25 to commence work on penicillin, Florey sat with Fleming and representatives of government, industry, science and the military in the offices of the Ministry of Supply as Sir Cecil Weir, the permanent head of the department, uttered words Florey thought he might never hear: 'Gentlemen, the Government will provide whatever financial means are required. All available knowledge and expertise will be pooled so that this drug can be produced without delay on a factory scale. Nothing will be allowed to stand in the way'.

While Florey's years of quiet pleading had failed to spark official interest and backing, the foghorn of Fleet Street could not be ignored.

« • »

Despite losing the battle for popular acclaim, Florey went on to win where he thought it counted more: among his scientific peers.

As British historian of medicine, Bill Bynum, said of Fleming's and Florey's respective contributions, 'The discovery was old science, but the drug itself required new ways of doing science'. Fleming did possess the 'prepared mind' necessary to pounce on his chance observation of penicillin mould. But the years spent bent over the bench in his little lab at St Mary's 'play[ing] with microbes' were of no help in turning the observation into invention. His eureka moment proved to be another scientific cul-de-sac.

It is, after all, one thing to identify an antibacterial mould, but Florey took on another, much more difficult task: to turn the mould into a new drug. 'Howard Florey can ... be regarded as a model for the contemporary biomedical scientist', wrote a fellow Australian winner of the Nobel prize for medicine, Peter Doherty. Unlike Fleming, Florey was a pioneer and an early master of 'new ways of doing science'. From the beginning of his career he had developed strong professional networks, including John Fulton and Alfred Richards in the United States; at the Dunn School he built a versatile team, bringing in the best biochemists he could find, Chain and Heatley; he gambled on penicillin mould; raised the money; isolated the active ingredient and produced penicillin the drug; tested it on mice and then humans and found it worked; knew he could not make enough in Britain so flew to America, and with the help of his network, convinced government and pharmaceutical companies to produce it on an industrial scale, sparking a medical revolution. 'What happened then is a paradigm for what happens today', concludes Doherty.

'Florey was not a profound visionary', Sir Henry Harris, a fellow Australian who succeeded Florey as Director of the Dunn School, summed up his former teacher and colleague. 'He was not a towering scientific intellect like Newton or Darwin or Einstein. But he had one supreme virtue: he knew exactly what had to be done next, and he got it done'. This was what mattered above all else. Florey had the courage and 'commitment in the face of uncertainty' (the quality that Andrew Hargadon sees as 'the defining characteristic of innovation') to persevere through shadows towards the light. A long succession of professional laurels, not least the Nobel prize, bestowed upon Florey by his colleagues recognised that crucial aspect of his scientific temperament.

If, by early 1943, the penicillin challenge seemed to have been solved, for Florey it was not yet the end of the quest, though perhaps the beginning of the end. The new drug would now take him a world away from the familiar corridors of the Dunn School and the streets of New York, his footsteps soon treading the sands of the Sahara.

《 • 》

With his eye firmly set on using penicillin to treat injuries of war, in July 1942 Florey sent a few minute samples of penicillin to the front in Libya. There was so little available he issued instructions that it be for surface use only. Even so, penicillin's debut in a war zone was successful enough to encourage Lieutenant Colonel Robert Pulvertaft, a Cambridge-trained pathologist, to grow his own mould and make penicillin juice, which he then applied directly to soldiers' wounds. By autumn, after the battle of El Alamein, more penicillin arrived from the Imperial Chemical Industries labs and, late in 1942, Florey himself prepared to go to North Africa.

After delays he feared interminable, Florey finally reached Algiers in early May 1943, having avoided a military rank and uniform that the British armed forces seemed determined to award him. His old friend from student days at Oxford, Hugh Cairns, joined him, and the two Australians shuttled around desert hospitals treating the Allied wounded and determining how best to use penicillin in the field.

In Tunisia, Florey and Cairns encountered a tricky issue. For Florey, treating seriously injured soldiers was one thing, soldiers who had caught sexually transmitted infections quite another. Though not especially prudish, Florey was less keen to use precious supplies of penicillin on self-inflicted miseries. Yet venereal diseases were keeping hundreds, if not thousands away from the front, including some of the most highly trained troops, like paratroopers.

The matter eventually reached the prime minister's office. Churchill, loath to decide whether to first treat those wounded 'on the battlefield or the bordello', replied cryptically, 'This valuable drug must on no account be wasted. It must be used to the best military advantage'.

In the contest between fairness and utility, utility won. Penicillin could cure gonorrhoea within just 48 hours, allowing paratroopers a descent from the heavens pain-free and ready for action. 'The goal', as a US War Production Board officer whimsically put it, now became 'to make penicillin so cheaply that it costs less to cure [a sexually transmitted disease] than to get it!' The goal was soon achieved, the price falling from $20 per dose in 1943 – $9000 per ounce or 250 times the price of gold – to 65 cents by 1945. The official history of the OSRD concluded, 'Penicillin alone was worth more than an Army corps in speeding recoveries for syphilis, gonorrhea and pneumonia'.

By the time of the Allied invasion of Sicily on 10 July 1943, troops had access to some but not nearly sufficient penicillin. Having trained hundreds of medical personnel in the best use of the new drug, Florey arrived back in Britain by air on 1 September. Two days later, General Sir Harold Alexander's 15th Army Group invaded mainland Italy. More penicillin was becoming available, enough for the most needy casualties. Two out of every three cases of gas gangrene now survived, thanks to the new drug. 'We've snatched them right out the grave', commented one amazed American doctor. By late 1943, with production methods in constant and dramatic improvement, American and British factories were working 24 hours a day to make sure that when the long-awaited D-day finally came, there would be enough penicillin for all casualties, serious and mild, bordello and battlefield, without the need for doctors and military brass to make painful choices. For the US War Department, only the atomic bomb had higher priority.

By now, Florey's work as a penicillin scientist and salesman was largely done. He was ambassador-at-large for the antibiotic revolution. Just before Christmas 1943, Florey left Oxford for Moscow to introduce the miracle drug to the Soviets. He got sick on the way, and, along with other delays, spent nearly four weeks in Tehran, before finally arriving in Moscow in late January. Two years before, the German army had stood at the approaches to the Soviet capital, the golden domes of the Kremlin glistening in the binoculars of Wehrmacht scouts; now Russian soldiers had pushed the enemy back to the Belarussian border. Despite shortages, the Soviets made Florey and his colleagues welcome, sparing nothing, not even hot water for his hotel.

《 • 》

Roosevelt, Churchill and Stalin met in Tehran in late November 1943, a month before Florey's unscheduled stop there to recuperate, for their first face-to-face summit as Allied leaders. The war in Europe would go on for another 18 months but the Big Three could smell victory. After agreeing to the invasion of Normandy in May 1944, Churchill became sick on the trip home with pneumonia. The penicillin myth was already potent in the public imagination when newspapers proclaimed the prime minister had been cured by Fleming's 'invention'. This was wrong on two accounts; Churchill recovered using standard sulpha drugs, provided by none other than Lieutenant Colonel Pulvertaft, North Africa's first cultivator of penicillin mould.

But, intriguingly, it might have been a different story for his arch-antagonist.

Florey was aware of German interest in penicillin as early as April 1941, shortly before his trip to the United States. He did all he could to ensure that the mould did not fall into Nazi hands. When CIBA, a Swiss company, contacted him to request *Penicillium notatum*, Florey did not respond, instead alerting the authorities, which scrambled to track down the whereabouts of all samples of the magic mould. 'During the past 10 years I have sent out a very large number of cultures of Penicillium to all sorts of places', Fleming told them, 'but as far as I can remember none have gone to Germany'.

But with the Third Reich in control of nearly all of Europe, it didn't really matter. German scientists soon demanded the Pasteur Institute in France as well as Dutch researchers hand over their supplies of *P. notatum*. Both sabotaged the German effort by providing strains that did not produce penicillin. The Dutch even made penicillin right under the nose of the German occupiers in the Netherlands Yeast and Spirit Factory in Delft; the German guard was kept drunk on gin to keep him in the dark. The Nazis did not succeed in manufacturing penicillin until October 1944, and then only in tiny quantities. Allied air raids crippled any possibility of mass production of the drug.

But if German soldiers could expect no miracles, there is an intriguing possibility that Florey's penicillin saved the life of Adolf Hitler. By the autumn of 1943, the new drug was making its way into Germany through neutral countries and, in small amounts, from captured Allied airmen who carried it for their own emergency use. Hitler's doctor, Theodore Morell, had access to this precious penicillin. Described by Albert Speer as 'a screwball only interested in money', Morell had an eye for the main chance. Aware of its miraculous healing powers, he reportedly used penicillin to treat Hitler's chronic throat infections and, momentously, following the attempt on Hitler's life on 20 July 1944 at his Wolf's Lair field headquarters in East Prussia. The bomb left by Claus von Stauffenberg blasted wooden splinters into Hitler's legs, which might easily have become infected. Morell's treatment with penicillin 'arguably did save Hitler's life'.

«•»

Arriving back in Oxford on 29 March 1944, exhausted after spending six of the last twelve months travelling in North Africa and Russia, Florey expected a well-deserved break. But then his country called.

Australia's prime minister, John Curtin, was in London in May 1944 attending an Imperial Conference and asked Florey if he would visit Australia to advise on penicillin. Florey agreed. Learning that the man responsible for making the new wonder drug was on his way home, the Americans also wanted him to lend a hand in the Pacific theatre of war. With US forces progressively island hopping to within bombing range of Japan, Japanese defenders were fighting with ever greater desperation and ferocity. On the tropical jungle battlefields, infection was a soldier's constant companion.

Florey travelled via the United States to American field hospitals in the Pacific and then the townships of Madang, Alexishafen and Lae on the north coast of New Guinea. The battle-hardened medics enjoyed this unexpected opportunity to host the world's leading authority on penicillin, as he examined rows of wounded warriors, many of whom would have been dead but for his miracle new drug. Florey was delighted, as were the soldiers and their doctors: even in the toughest conditions, penicillin was saving lives and limbs. 'After fighting disease and battle infections with sulfa drugs the cry became, "praise the Lord", when penicillin became available', recalled one army medic.

In late August, Florey returned to Australia on an American bomber, flying over Darwin, still scarred by Japanese air attacks. Recently knighted by King George, Sir Howard arrived to a hero's welcome from a grateful nation only now daring to believe that the Japanese threat was abating.

Florey was heartened to learn that his compatriots were at the forefront of penicillin production. Two enterprising Australians, Percival Bazeley and HH Kretchmar, had travelled to Peoria and brought back to Australia a culture of Mary Hunt's high-yielding penicillin. Australia had started independently manufacturing cavity magnetrons and microwave radars a year before, and now the Commonwealth Serum Laboratories in Melbourne began making penicillin. By 1944 they were producing enough not only for Australian soldiers in the Pacific theatre, but in a world first, enough for civilians as well, with some left over for American brothers-in-arms (the first civilian treated by penicillin in Australia was 7-year-old Peter Harrison, whose life-saving doses were flown by a Liberator bomber from San Francisco, through Hawaii, to Sydney in July 1943).

Proud to be back in Australia, honoured and successful, Florey was now a public figure. But a reluctant one. Even in his homeland he tried to avoid the press – albeit not very successfully – and concentrated on lecturing doctors about penicillin best practice. In late September he returned to Adelaide, where he reunited with two of his sisters, Hilda and Valetta, and received the degree of MD from the University of Adelaide.

In accepting the award Florey traced his own story and sought to pave an easier path for those that followed:

> The time has come to mention whether Australia can afford to lose its bright young people; whether it is not time to take a leap forward in scientific research ... I have the greatest confidence that the leap can be taken here, and that, within a few years, Australia will have centres that compare favourably with those overseas.

This cause would soon capture Florey's imagination. Before departing Australia, Dr HC 'Nugget' Coombs, the head of the Department of Post-War Reconstruction, invited Florey to write a memorandum to the prime minister, John Curtin, giving his views on establishing just such a centre of excellence in medical research, to be based in Canberra. It was the beginning of a mission that would loom large for the rest of Florey's life.

The man who boarded an American bomber in early October to return to Oxford was no longer the enthusiastic and energetic young scientist who had set out from Adelaide more than 20 years before. The last five years had taken their toll on Florey. He was exhausted. But he knew he had changed medicine and human health. It was a race well run.

《 • 》

Florey failed to keep penicillin secret from the public until its safety and supply could be guaranteed. In the space of a few short months in mid-1942, the drug had become a sensation for grateful soldiers while civilians eagerly anticipated its homefront debut. With 'penicillin' entering everyday English as 'atomic bomb' and 'radar' would shortly, the miracle new treatment was now in the public spotlight.

Fleming was not the only one to claim paternity. To his sister Hilda, Florey wrote of 'the spectacle of everybody who has ever had the slightest bit to do with it crowding in to prove how important their contributions to the matter are'. In the United States, Dr Rene Dubos was for a while touted as the source of the Oxford group's interest in penicillin. While that claim was quickly put to bed, others were more contentious. The Rockefeller Foundation, in their 1944 Annual Report, claimed much of the credit for development of the drug, forcing Sir Edward Mellanby and the Medical Research Council, as well as the British government, to publish just how much (or how little) they had contributed to the enterprise. For good measure, Dr Andrew Moyer, Heatley's co-worker at the NRRL in Peoria, basked in the glory of his foreign patents on new and improved methods of penicillin production.

Florey tried to stay above the fray, refusing to become involved in the public brawl for credit for the scientific discovery of the century. But it hurt. Florey bristled with indignation at the lack of recognition for his Oxford team. Mellanby sought to soothe the man he had chosen to back in 1939, telling him:

> In time, even the public will realize that … the thing that has mattered most has been the persistent and highly meritorious work of your laboratory. The dish you have turned out is so good that you must swallow the rather nauseating but temporary publicity ingredient with a smile.

Mellanby's words have not aged well. More than 75 years later, Fleming remains popularly credited with starting the antibiotic revolution. To Google 'penicillin' is to retrieve the myth afresh. Legends die hard; some are indeed immortal.

The 'rather nauseating ... publicity' had immediate implications that, but for some timely interventions from friends and colleagues, would have proven all but impossible to swallow. Award of the Nobel prize had been suspended during the war, and anticipation was keen as to who would win in the category of Physiology or Medicine for 1945. There was little doubt that penicillin, by now recognised globally as a wonder drug, would be the subject of the award, but much lobbying and speculation surrounded the question of who would be given credit and awarded the prize.

For a good while both Florey and Chain believed they might be overlooked. Unable by temperament and conviction to lobby for recognition, they feigned that they did not care. But they did. Fortunately, so did many others in the world of science who knew the real story of penicillin and understood the crucial difference between 'discovering a mould' and 'developing a drug'. They picked up the phone, wrote letters, and sent telegrams to make sure Florey and his team received their due.

First among them was John Fulton. In early October 1944, he wrote to the Swedish legation in New York representing, he said, 'the considered opinion of a large group in this country including the Division of Medical Sciences of the National Research Council and the officers of the American Medical Association'.

'Our recommendation', Fulton was quite explicit, 'would be, in order of preference:

'Florey alone
'Florey and Chain
'Florey, Chain and Fleming'.

For good measure, he followed up a couple of weeks later with a cable, all in shouting capital letters, to Dr Göran Liljestrand, Secretary of the Nobel Committee: 'PROFOUNDLY DISTURBED OVER PERSISTENT REPORTS THROUGH BEAVERBROOK PRESS CONCERNING FLEMING SINCERELY HOPE RUMOUR IS UNFOUNDED SINCE CREDIT FOR ALL CLINICAL DEVELOPMENT BELONGS TO FLOREY AND ASSOCIATES'.

Lobbying continued for another 12 months, with opinions see-sawing among decision makers in Stockholm. Sir Henry Dale pitched in for Florey, along with others who could not be ignored. In the end, nearly all scientists urged inclusion of Florey and Chain. Without the two Oxford scientists, the considered professional opinion went, penicillin would have remained just a funny little mould, promising but mostly forgotten, and not a powerful new drug available on the beaches of Normandy and beyond.

The Nobel Committee agreed, and on 25 October 1945 Florey received the telegram. He read that along with Fleming and Chain, he had been awarded the Nobel prize 'for the discovery of penicillin and its curative action in various infectious diseases'. For Fleming, it crowned a two-year campaign to rescue himself from scientific obscurity;

for Chain, it represented well-deserved recognition of a biochemist's crucial role in the process of discovery and development; for Heatley, perhaps, a disappointment he accepted with his usual grace; but for Florey it was the final and most important vindication. The six-year marathon from a risky Petri dish bet to a life-saving phenomenon for the world was run, won and now finally done, the race truly for the ages.

13

Critical mass

Oliphant returned to Birmingham from America in mid-October 1941. There was no hero's welcome and a dreary Midlands autumn awaited him.

Truth be told, Oliphant did not yet know if his heroic efforts had paid off. He wrote to Lawrence just before he left, 'I feel quite sure that in your hands the uranium question will receive proper and complete attention, and I do hope that you are able to do something in the matter'.

It did and he was, but neither Oliphant nor the British government would hear about it for months. Like Florey, Oliphant barnstormed around the continent trying to stir up interest, but too many seemed unmoved and unconvinced. Finally, again like Florey, Oliphant had to stake everything on an old friend's offer of help. It was a tenuous lifeline. There was nothing left but hope.

As he worried and waited for news from America, Oliphant was humiliated by another development. The

MAUD Committee, 'one of the most effective scientific committees that had ever existed', was quietly disbanded. There was no thanks for its groundbreaking work, not even an official response to its final report. Even worse, the new committee responsible for the bomb project, codenamed Tube Alloys, was headed by two executives from Imperial Chemical Industries whom Oliphant considered interlopers.

He was outraged. The end of MAUD was bad enough, but for its successor body to be run by a brace of businessman was the final straw. Mustering his indignation, Oliphant wrote to the head of the Department of Scientific and Industrial Research, Sir Edward Appleton, who oversaw Tube Alloys. Both Oliphant and Appleton had conducted research under Ernest Rutherford at Cambridge. Such connections, however, did not dampen the ferocity of Oliphant's letter to him:

> I can see no reason why the people put in charge of this work should be commercial representatives completely ignorant of the essential nuclear physics upon which the whole thing is based ... In the reorganisation [in early 1941] I was left off the policy committee and I feel that the time has now come for me to sever my connection altogether. This problem is too important to be trifled with.

Hoping to soothe a figure so central to the birth of the atomic-bomb project, Appleton invited Oliphant to London for a private briefing. Oliphant emerged mollified but still sceptical of industry's designs on exploiting nuclear research work after the war. He remained on the newly constituted Tube Alloys technical committee, but he was no longer at the heart of the action. He might have enjoyed train trips to London with fellow committee member Rudolf Peierls, but politicians, administrators and company men were taking over. Oliphant comforted himself by returning to radar work. He was soon so busy in his old research fiefdom he had little time for Tube Alloys.

And then Australia called. Japanese armed forces were now uncomfortably close to home. They invaded New Guinea in January 1942 and bombed Darwin in February. Worried about Australia's precarious defence, Oliphant offered to help with developing microwave radar for use in the Pacific theatre. The Australian Government accepted and in March 1942 he left Britain. It suited his British superiors to let go the abrasive, indiscreet Australian who was no longer vital to developing Britain's top-secret work. He was now out of the loop and dismayed with bureaucrats. But Oliphant need not have worried about the future of the nuclear program. By the start of 1942 the Americans had wholeheartedly committed to building the atomic bomb.

« • »

Oliphant later learned about the chain reaction he had started in the United States. His time spent with Lawrence at Berkeley had led to Lawrence's meeting with Compton and Conant in Chicago in late September 1941, which in turn led to Bush's briefing of the president in early October and a third review of uranium research from the National Academy of Sciences delivered in November. The final go ahead from the White House came the day before Pearl Harbor. The venture was subsequently named the Manhattan Project after the location of its first headquarters. By the time Oliphant arrived in Australia to lend his hand with radar research and production, the Americans had taken the first concrete steps on the road to Hiroshima and Nagasaki.

Most importantly, after the Japanese attack on America, Roosevelt and an outraged US Congress finally made funds available to do the work of war – money beyond the wildest dreams of Lyman Briggs and his committee, doling out a few thousand dollars to lonely and undirected physicists to paint the sky blue. The United States moved silently and swiftly towards the bomb as if it had always been the plan. The British origins of the project were largely brushed aside and lost among the clamour of activity that erupted in early 1942.

Churchill was at least partly responsible for the sidelining of Britain. He missed the small window of opportunity when formal scientific cooperation between the two countries could have been secured on more equal terms. A couple of days after the fateful White House briefing by Bush in October 1941, Roosevelt wrote to the British prime minister proposing a partnership: 'It appears desirable that we should soon correspond or converse concerning the subject which is under study by your MAUD Committee and by Dr Bush's organisation in this country in order that any extended efforts may be co-ordinated or even jointly conducted'.

Instead of responding, Churchill stalled.

While Britain had shared the MAUD report and other documents to spur America into action, it feared that closer entanglement would endanger its lead in nuclear science. This was hubris and a folly, for without the necessary money and manpower Britain soon fell behind America in the atomic race. Even before Oliphant's trip to America, Sir John Anderson, as the minister responsible for Tube Alloys, advised Churchill,

> We must, however, face the fact that the pioneering work done in this country is a dwindling asset and that, unless we capitalise it quickly, we shall be outstripped. We now have a real contribution to make to a "merger" [with the United States]. Soon we shall have little or none.

And so it came to pass. Churchill dithered; the Americans wasted no time. While no one knew how long it would take to turn Frisch and Peierls' calculations into a workable weapon, the best bet was three or four years. Everyone knew it would require a massive effort and commitment of resources. The National Academy of Science's third and final report to Vannevar Bush estimated that the atomic-bomb project would cost about $133 million. American science administrators knew it would likely be much more, but no one, Bush most of all, wanted cost estimates to foil the project. As the expenses increased exponentially, the savvy 'science czar' ensured that the atomic bomb budget was buried within the enormous yearly appropriations of the Army Corps of Engineers. The cost of the Manhattan Project – totalling over $2 billion, or about 15 times the original estimate – was met without ever having to seek direct Congressional approval.

As more than two dozen new industrial and scientific sites sprang up across the country – the majority tasked with producing sufficient fissile material for the bomb – employing more than 130 000 scientists, technicians and other workers, it was soon too late for Britain to join as an equal partner. While many British scientists would end up working on the bomb, the United States remained in full charge of the project.

In time, Oliphant got to America too. His return to Australia did not work out as planned. Frustrations plagued him from the start. He could not find a seat on any aircraft and the voyage back home took ten long weeks by ship. When he finally arrived, he was not made welcome. For reasons that are not clear, the leader of the team that had invented the cavity magnetron and revolutionised modern warfare was not required for Australia's radar program. He was disheartened by his inability to make a worthwhile contribution and, reunions with Rosa and the children aside, soon wished he had never left Birmingham.

Before long, Oliphant sought to return and booked passage for his family on *Desirade*, a rough-looking French vessel bound for Durban. And that's as far as they got, for the ship named Desire was deemed unfit for further passenger travel. When they had to wait a month for another boat, the Oliphants made the most of it and celebrated Christmas 1942 in nearby foothills. The family arrived back in Britain at the end of February. Frustrated at having wasted 11 months on an ill-fated patriotic odyssey, Oliphant moved quickly to re-establish himself in the thick of things.

At first, he threw himself into further refining and manufacturing the cavity magnetron, the technology that was at that very moment turning the tide of war in the mid-Atlantic. But his mind soon wandered back to nuclear physics and the atomic bomb. Timing was on Oliphant's side. In mid-August 1943, Roosevelt and Churchill met secretly in Quebec and, spurred by the slower than expected progress so far, agreed that from now on there must be full collaboration between the two countries to build the atomic bomb. As the British pulled together their top physicists for the job, including Frisch and Peierls, Oliphant's claim could not be ignored. If he was willing to give up the Admiralty's radar work, he would be welcomed back among the scientific brotherhood of the bomb.

As it happened, it suited Oliphant, believing others could now carry on with radar. He arrived in the United States in September 1943 at almost the exact mid-point of the wartime atomic saga: two years after his first trip to shake America from its complacency and just under two years before the bombs were dropped on Hiroshima and Nagasaki. After Quebec, the Manhattan Project swallowed Tube Alloys whole. Oliphant, who had a claim to be the godfather of both programs, was back in the hunt for the bomb. He had one more scientific contribution to make.

《 • 》

In Washington, General Groves and Oppenheimer met with Oliphant to entice him to Los Alamos. Groves was worried that Oliphant and Lawrence were too friendly and showed little discretion. But Oliphant waived away Groves' overtures. He was keen to work with Lawrence on what he regarded as the most critical task: separating uranium isotopes. The bomb needed uranium-235, regardless of all the physics and engineering brilliance in the New Mexico desert. Oliphant and his small troupe of six or so from Birmingham joined Lawrence and his team of around a thousand scientists to isolate and harvest the readily fissionable but rare uranium-235 from the bulk of the stable and useless uranium-238. Oliphant and Lawrence bet on the electromagnetic method, rejecting gaseous diffusion that Oliphant regarded as ponderous and problematic. The former was a no-brainer for the two nuclear physicists, since the process would use their treasured cyclotron.

There is a brief hint Oliphant met someone else in Washington beside Groves and Oppenheimer. In a story he told his grandson, Oliphant claimed that he was also called to the White House to reassure the president about the ultimate prospects of the bomb project. Roosevelt continued to be concerned about the slow grind of the Manhattan Project. Oliphant, an outspoken outsider unbeholden to the American scientific and military establishments, was arguably the best person to convince the commander-in-chief that technical problems hampering progress were surmountable and that the bomb would be built in time to play its part in this war. While not officially reported, such top-level intervention would have been up Oliphant's alley, and consistent with his past passionate advocacy for both the viability and the necessity of the atom-bomb project.

At Berkeley, Oliphant was back in his element. He was excited by the challenges and, if no longer a central figure, remained an important player, formally number two in the British chain of command behind James Chadwick. For Oliphant, the next 18 months would be among the most stimulating in his professional life, rivalled only by the wonderment and comradeship he felt in his salad days at Rutherford's shoulder at the Cavendish Laboratory. Though now in his early 40s, Oliphant's energy, enthusiasm and idealism never faltered. He would commonly work half the night powered only by coffee and cigarettes. He did not care and never complained. His colleagues loved him for it.

Oliphant and Lawrence's work commenced at Berkeley but soon included a huge industrial plant in Oak Ridge, Tennessee, wholly dedicated to isotope separation by electromagnetic means. The Oak Ridge facility used more electricity than all of Canada and the entire US stockpile of silver for its magnets in lieu of copper wire needed elsewhere. But even with all the resources that the United States could muster, turning an experimental laboratory process into the basis for industrial-scale production of fissionable material took ingenuity and hard work. The 'Lawrence–Oliphant electromagnetic method' involved long belts of giant magnets, called 'racetracks', pulling apart the isotopes as gaseous uranium 'raced' through vacuum chambers sandwiched in between. When a worker carrying a metal sheet found himself stuck to the racetrack, he had to be laboriously pried off with planks of wood, since to shut down the machinery would mean losing a day's worth of output.

In the end, both electromagnetic separation and gaseous diffusion – scorned by Oliphant but improved by others – supplied, milligram by milligram, uranium-235 to the bomb's technicians in Los Alamos. Plutonium, element 94, emerged as another option to provide the bomb's critical mass. Both were used to create weapons.

By the beginning of 1945, Oliphant knew he was no longer indispensable. The work at Berkeley and Oak Ridge was nearly finished and success all but assured. He had enjoyed his time with 'livewire' Lawrence and his team and would stay in close touch until Lawrence's early death in 1958 at the age of 57. But his thoughts now drifted to work and life after the war. He wanted to establish the cyclotron at Birmingham and refocus his department's research after years of war work. And he missed his family too.

He lingered in Berkeley at Lawrence's urgings but in April Oliphant officially ceased work on the bomb. He had done his bit. The war in Europe was coming to an end: Hitler would be dead within the month, and Germany formally surrendered on 8 May 1945. There was no last-minute reprieve for Germany with a feared Nazi super-bomb though it turned out this was not for lack of trying.

《 • 》

'A hatred of the Hitler regime' drove Oliphant to convince the British and American governments to commit to atomic-bomb programs. It gave those employed on Tube Alloys and the Manhattan Project – at least those with security clearance to know what they were working on – the motivation to get there first. The prospect of Hitler's nuclear bomb terrified Oliphant, as it did Frisch and Peierls. Hindsight is a luxury not available to contemporaries.

After all, uranium fission had been discovered in Germany. Despite the exodus of Jewish scientists, Germany boasted some of the greatest nuclear physicists in the world, few with higher standing than Werner Heisenberg. 'We must make use of physics for warfare', proclaimed a Nazi slogan, which Heisenberg reworded as 'We must make use of warfare for physics'. It is debatable whether physics or warfare got more out of the other under Hitler, but after meeting with Heisenberg in Copenhagen in September 1941, Niels Bohr was left in no doubt that Germany was doing all in its power to build nuclear weapons. Physicists in

Britain and later the United States assumed that in the race for an atom bomb, Germany enjoyed a head start.

Unbeknown to the Allies, it also had the necessary materials, with a plentiful supply of uranium from the mines of Joachimsthal in the Sudetenland. Otto Hahn and other German nuclear scientists were using it as fuel for their uranium piles as late as 1945. After the area was returned to Czechoslovakia, the Soviets made German POWs excavate the uranium, and used it to make their first atomic bombs in 1949. The Soviet Union had many talented scientists, but when communist spies at Los Alamos and elsewhere stole the blueprints of the bomb, they had no need to reinvent the wheel.

The explanation most often given for why Nazi Germany never developed a uranium bomb is that the research, badly mismanaged by competing agencies and priorities, came to a dead end. Heisenberg claimed after the war to be shocked when he learned the details of the physics and engineering behind the American bomb, telling his Allied interrogators he and his colleagues never contemplated such a thing. Heisenberg's protestations can safely be dismissed as myth. He later told a historian that before the Americans had ever committed to building it, 'we saw an open road ahead of us, leading to the atomic bomb'. German physicists were no fools. But technical know-how and expertise were not the only ingredients necessary for success.

Another view is that while German physicists were confident that a fission bomb could be made, the Nazi hierarchy did not pursue it, believing the war would be over before it was done. Then, as the war lingered on, it was too late. The massive resources needed were unavailable, swallowed by numerous other projects, such as ballistic rockets, jet fighters and a new generation of U-boats. In other words, the decision not to go all out for a bomb was a political one – not a failure of science.

Yet another theory has German atomic scientists prevaricating, dissembling and deflecting the prospect of a workable nuclear device either because they feared a failure that would not be tolerated by their political masters, or from furtive opposition to the Nazi regime. 'I would definitely have considered making atomic bombs for Hitler a crime', Heisenberg wrote to the Dutch mathematician Bartel Leendert van der Waerden in 1948. Such sentiments were common in postwar Germany, and often smacked of self-serving historical revisionism, yet there is some contemporary evidence to support them. On 15 April 1941, Lyman Briggs received a note from Rudolf Ladenburg, a German-born physicist at Princeton University, stating:

> It may interest you that a colleague of mine who arrived from Berlin via Lisbon a few days ago, brought the following message: a reliable colleague who is working at a technical research laboratory asked him to let us know that a large number of German physicists are working intensively on the problem of the uranium bomb under direction of Heisenberg, that Heisenberg himself tries

to delay the work as much as possible, fearing catastrophic results of a success. But he cannot help fulfilling the orders given to him, and if the problem can be solved, it will be solved probably in the near future. So he gave the advice to us to hurry up if U.S.A. will not come too late.

There is no evidence that Briggs was interested. Like the MAUD documents, Ladenburg's alarming note did not spur him into action.

Whether politics or physics were ultimately to blame, the German atom-bomb project made little progress. It did not help that Allied saboteurs in Norway destroyed a fresh stockpile of 'heavy water', which German scientists thought essential to making a bomb. By mid-June 1942, Albert Speer advised Hitler that while the nuclear program might be useful in the long term, there was no decisive super-weapon on the horizon. Hitler and his Armaments Minister turned their attention elsewhere, as consequently did Heisenberg and his colleagues. However, when radar scientist RV Jones, at that time monitoring German scientific publications for British intelligence, noticed that, as he exclaimed, 'the German nuclear scientists have stopped publishing!' the alarmed War Cabinet met within an hour to consider the implications. What Jones and the British government took as an indication the Nazis had commenced building a bomb in fact signalled a loss of interest. With no certain way to find out, the fear of a Nazi bomb never left the Allies.

Japan also entertained the possibility of an atomic bomb. In October 1940, more than a year before Pearl Harbor, Lieutenant General Takeo Yasuda of the Imperial Japanese Army read a report that concluded that it would be possible to build an atomic weapon. Yasuda contacted Yoshio Nishina to see if he could assist. Later dubbed Japan's 'father of nuclear physics', Nishina had studied under Niels Bohr and was a friend of Ernest Lawrence. With Lawrence's help in 1937 he built the first cyclotron outside the United States and subsequently built another four. Nishina's lab was tasked with researching an atomic bomb in April 1941 and concluded that a bomb was theoretically but not technically feasible. In 1943, lacking resources, including uranium, to build their own, Nishina and the Japanese Government, concluded that while the Americans were probably working on a bomb, not even they could build one in the foreseeable future.

《 • 》

In the desert, dawn broke brighter than a thousand suns.

At 5.30 am on 16 July 1945, the Americans detonated the world's first atomic bomb 300 kilometres south of Los Alamos, at the Trinity test site near Alamogordo in New Mexico.

Otto Frisch, who was there, described the historic moment:

And then, without a sound, the sun was shining or so it looked. ... I kept blinking and trying to take looks, and after another ten seconds or so it had grown and dimmed into something more like a huge oil fire … And all in complete silence; the bang came minutes later, quite loud though I had plugged my ears, and followed by a long rumble like heavy traffic very far away. I can still hear it.

Peierls, also present to witness the awesome physical demonstration of his early pencilled calculations, remembered the reaction: 'We were struck with awe. We had known what to expect, but no amount of imagination could have given us a taste of the real thing'. Oppenheimer, touched and traumatised by the spectacle, resorted to quoting the Hindu holy book, the Bhagavad Gita: 'the radiance of a thousand suns … I am death, the destroyer of worlds'.

Nicknamed 'Gadget', the plutonium bomb detonated atop a 30-metre steel tower and generated a fireball with a temperature four times that of the core of the sun. Its shockwave indicated a blast of 18 600 tons of TNT. The Earth trembled as trillions and trillions of uranium nuclei shattered in an unstoppable chain reaction, a powerful heat blast fusing sand into glass and turning shrubs into vapour.

General Groves, Vannevar Bush and James Conant were also at the Trinity site. They congratulated each other with silent handshakes, knowing that the war might soon be over. The bomb came too late for Germany, which had capitulated two months earlier, but Japan was still fighting.

Just four hours later, a uranium bomb, named 'Little Boy', left San Francisco harbour onboard a navy vessel bound for the Mariana Islands. The world would never be the same.

《 • 》

The 'silver-plated' B-29, named by its commander, Colonel Paul Tibbets, *Enola Gay* after his mother, took off into the dark at 2:45 am on 6 August 1945.

The Superfortress was several tons overweight and needed every inch of the 2-mile runway on the small island of Tinian in the northern Pacific archipelago of the Marianas to take off. Squeezed into the bomb bay and looking like 'an elongated trash can with fins' was the 4-ton 'Little Boy'.

Shortly after 5 am, the first light of dawn appeared in the east, and an hour later, from 9300 feet above, the crew could see the island of Iwo Jima, a recent bloody battlefield that cost Americans more than a thousand casualties for each square kilometre of its volcanic ashen ground.

Less than three hours from the Japanese home islands, Tibbets was still not sure which city he would target. It depended on the weather. Then came confirmation: it was clear over Hiroshima, home of Japan's Second General Army and an important embarkation point for imperial troops.

The bomb was dropped exactly 17 seconds after 8:15 am local time, exploding 43 seconds later at 1900 feet above the city with a power equivalent to 12 500 tons of TNT. The flash of 60-million-degree heat was followed by the shockwave travelling faster than the speed of sound. Some

70000 to 80000 people died almost instantly. About the same number more would be dead by the end of the year from radiation, burns and other injuries. The mission was a military success, and a great human tragedy.

A dozen crew were onboard the *Enola Gay* that morning. They included a weaponeer and assistant weaponeer to arm the bomb, and a bombardier to drop it; as well as a radar operator, a radar countermeasures officer and a navigator, who also relied on radar. Nearly half the crew were working with technologies that had not existed just a few years before: airborne radar and an airborne atomic bomb.

The B-29 was the first American bomber aircraft to use radar as standard equipment, and the *Enola Gay* also had AN/APQ-13 radar capable of accurately scanning the ground. It was a big improvement on Britain's early 10-centimetre H2S radar and a valuable refinement on the United States' standard 3-centimetre H2X radar. With it, high altitude, or 'blind bombing', became much more accurate as did navigation.

Tibbets recalled, 'Our orders were for a visual bombing run. Even though radar sighting had proved accurate throughout the war, it had been decided that this bomb should be dropped only if the target was clearly visible'. The blast was going to be filmed and so blue sky proved Hiroshima's death warrant. There is little doubt that *Enola Gay*'s onboard radar would have allowed 'Little Boy' to be dropped on a target obscured by clouds. In any case, the bomb came equipped with as many as four proximity fuzes to ensure it exploded at the predetermined altitude.

'Little Boy', the fruit of the third most expensive scientific-industrial project of the war (the Manhattan Project at $2.2 billion), was carried onboard a B-29 bomber (developed at a cost of $3 billion) and equipped with AN/APQ-13 radar and the bomb armed with proximity fuzes (microwave radar costing about $3 billion). The *Enola Gay* carried the most expensive and destructive hardware in history, much of it conceived in Mark Oliphant's laboratory in Birmingham.

On 9 August, three days after the bombing of Hiroshima, a second atomic bomb was dropped on the city of Nagasaki. It was a more powerful weapon, a plutonium bomb, but it was not delivered with the same accuracy and the city's hilly terrain inhibited the bomb's effectiveness. It did not matter. Nearly 40 000 Japanese were killed instantly, Nagasaki was devastated and the Japanese will to fight buckled. Coupled with the shock of the Soviet Union's declaration of war on Japan the previous day and the Red Army's immediate invasion of Japanese conquests in China and Korea, the Japanese surrendered on 15 August 1945. *Time* magazine, which was planning to put radar on the cover of its 20 August issue as the weapon that won the war, now devoted its pages to the atom bomb, relegating radar to page 72.

Without Oliphant's Birmingham lab, it is unlikely that microwave radar and the atomic bomb would have been developed in time to play a significant, history-changing role in the bloodiest conflict in human history. The war, nearly five years old, ended literally in a flash. But it came at a cost. Up until now only God had the ability to destroy the Earth. Now humans would also wield the terrifying power to end civilisation, if not life itself.

《 · 》

From the very moment the atomic bombs were dropped on Hiroshima and Nagasaki, their necessity and morality became perhaps the greatest controversy surrounding Allied conduct during the war.

The creators of the bombs were among the most conflicted. Oliphant himself did not leave a contemporary reflection but it seems almost certain that while he believed the bomb had to be built – just in case – it did not have to be dropped on an almost defeated nation. Returning from drinks in Los Alamos, Oppenheimer ran across a perfectly sober young physicist retching in the bushes at the thought of what he had helped achieve. Frisch subsequently wrote of his 'unease, indeed nausea', as it 'seemed rather ghoulish to celebrate the sudden death of a hundred thousand people, even if they were "enemies"'.

Since then, many have sympathised with the scientists' reticence. Perhaps the awesome destructive power of the bomb should have been shown to the Japanese over an uninhabited Pacific atoll. Or perhaps the bomb could have been used against a military target.

Beyond the question of morality looms the tactical consideration: was the nuclear double whammy at all necessary? Some have argued that Japan was already on the verge of surrender, with the end a matter of weeks if not days away. Historical and archival research, however, is equivocal in support for such an optimistic scenario.

Still others have pointed out that the Soviet entry into the war against Japan would have itself precipitated an early capitulation. But if the fate of the remaining Japanese armies in mainland Asia was sealed by Stalin's decision, doubts remain that the Red Army possessed the capacity to carry out a D-day–style contested seaborne landing on the home islands (the Kuril Islands were occupied only after the Japanese surrender on 15 August, by a small initial force, and with scant resistance). Even if facing a two-front invasion, there is no reason to think that Japanese soldiers and civilians would have fought any less doggedly against the Soviets than they had against the Americans. More so, perhaps, as the Japanese thought the Soviets more likely to kill the Royal Family and destroy Imperial Japan. The overall casualties would have probably been the same, though shared between the Red and the US Armies.

Some historians have suggested that the dropping of the atom bombs was primarily a warning to the Soviets as wartime amity gave way to concerns about Stalin's expansionist intentions. Perhaps it was that too, but contemporary records leave little doubt that the prospect of a protracted and bloody conquest of the Japanese archipelago haunted the decision makers and the troops alike. The Americans had had enough of fighting and dying.

If the Manhattan Project scientists had qualms about the weapon they helped to unleash, there is little doubt what most young American servicemen thought about the bombing of Hiroshima and Nagasaki. The GIs treated physicists in Tinian as heroes after the news about the new super-weapon was made public. Those whose job it would have been to land on the Japanese beaches and fight bitter resistance never had second thoughts. Historian Paul Fussell, at the time a 21-year-old rifle platoon leader wounded in France but scheduled for the invasion of Honshu, recalled, 'we broke down and cried with relief and joy. We were going to live. We were going to grow to adulthood after all'. And famed chronicler of the American Civil War, Shelby Foote, who was serving in the US Marine Corp observed: 'There's a lot of talk now about a guilt trip over dropping that bomb. Anybody who was in the army or the Marine Corp when they dropped that bomb never heard such hurrahs in his life'.

The island hopping towards Japan gave the American armed forces a bitter foretaste of things to come, and the closer to the home islands the more bitter it became. Clearing isolated Japanese garrisons that fought with suicidal determination to the last man and the last bullet was one thing; facing millions of imperial troops mingled with tens of millions of civilians in one of the most densely populated patches on Earth was an appalling prospect. The conquest of Okinawa made for a grim preview: 98 days of bloodshed, fought over a mere 1200 square kilometres (half the size of the Australian Capital Territory), costing the lives of around 15 000 Americans, 77 000 Japanese troops and up to 100 000 Japanese civilians, or a third of the island's population (a similar proportion to those who would later die in Hiroshima). Only 7000 Japanese soldiers surrendered.

Japan's plan for the defence of the homeland envisaged resistance by some two million troops, even if underarmed and underequipped, and a civilian defence force of 28 million, cannon-fodder often outfitted only with pitchforks and sticks. The official propaganda campaign launched in mid-1945 called for 'The Glorious Death of One Hundred Million'. It would have never come to that total – greater than the entire population of Japan at the time – but the American planners of Operation Downfall, the codename for the invasion of the home islands scheduled for early 1946, predicted up to a million American casualties alone. There were also 300 000 Allied POWs in Japan, under orders to be executed in the case of an invasion. As for the Japanese soldiers, irregulars and civilians caught in crossfire, estimates ran so high as to be almost abstract, with figures like 10 or 20 million difficult to comprehend and visualise, bearing in mind that Australia's entire population then stood at only seven million. Japan would have no doubt surrendered, but much later and knee-deep in blood. Then there were still several million Japanese troops in Korea, China, Malaya and scattered throughout the rest of Asia and the Pacific, who would have continued to fight.

From President Truman down to the lowliest private, most had come to believe and would continue to do so for the rest of their lives that for all the horror of the bomb, it most likely saved millions of both Allied and Japanese lives that would have been lost if Japan had been invaded. The atom bomb was terrible. The mass killing and destruction caused by a single explosive device were unprecedented in history. As were the long-term effects of radiation on survivors. The spectre of civilisation-ending nuclear Armageddon cast a lasting shadow over humanity. We are still living with its poisoned legacy. But for all its horror, the bomb most likely helped to end the war a year early (American military planners expected Japanese resistance to end in November 1946). Of the two Superfortresses that accompanied *Enola Gay* on its fateful flight to Hiroshima, the B-29 carrying photographic instruments was later named *Necessary Evil*. This, perhaps, is the closest we might come to the true understanding of that monstrous day.

Even after the dropping of the second, plutonium bomb on 9 August on Nagasaki, the war continued. Some in Japan questioned whether the bombs were really nuclear weapons and whether the Americans had any more to deploy (they did, but deployment was weeks away). The B-29s, meanwhile, kept coming, their last raid on 13 August on Tokyo involving 1500 bombers dropping their 1-ton payloads all day long with virtually no resistance.

With bickering among military and political leaders showing no signs of resolution, the Emperor finally stepped in. Perversely, the bomb had given him and his generals a way out. Asking his subjects to 'endur[e] the unendurable', he called a halt to hostilities, acknowledging, 'The enemy has begun to employ a new and most cruel bomb, the power of which to do damage is indeed incalculable'. The monarch's message was broadcast on the morning of 15 August 1945. It was the first time his voice had ever been heard on radio and they called it 'Gyokuon-hoso' or the 'jewelled sound'. After the flash of a thousand suns carried away two of his cities, the hushed and sombre tones of the Emperor of the Land of the Rising Sun, descendant of sun goddess Amaterasu, sounded an eerie coda in formal and florid Japanese to the six-year-long symphony of death and destruction. The war was finally over.

《 • 》

In assessments of the Second World War, it is a common conclusion that the atomic bomb ended the war, but microwave radar won it for the Allies. The director of the MIT's Rad Laboratory, Lee DuBridge, was only the first of many to make this point.

AP Rowe, who first blew the whistle on Britain's lack of preparedness for an airborne attack, later concluded that the cavity magnetron 'had a more decisive effect on the outcome of the war than any other single scientific device evolved during the war' and was 'of far more importance than the atomic bomb'. For some, this was already apparent at the time. So convinced was he of the pre-eminent importance of radar and his work at MIT's Rad Lab to the outcome of the war, future Nobel laureate Isidor Rabi turned down his friend Oppenheimer's offer of Associate Director at Los Alamos, telling him, 'I'm very serious about this war. We could lose it with insufficient radar'.

Vannevar Bush concluded, 'World War II was the first war in human history to be affected decisively by weapons unknown at the outbreak of hostilities'. In just a couple of weeks in early 1940, Oliphant's teams came up with the breakthroughs that made both these mega projects possible, first developing the cavity magnetron that formed the heart of microwave radar and, shortly after, providing proof of the feasibility of an airborne atomic bomb. Just over five years later, the device that helped to fight and win the war – radar – would accompany and direct the fateful flight to drop the weapon that ended it – the atomic bomb. And even as the fighting stopped, the third breakthrough, penicillin developed by Florey's team at the end of those crucial hundred days in 1940, would continue to work its miracles, so that millions more need not die.

Atomic bomb, microwave radar and penicillin were the three most significant inventions of the Second World War, brought to life in some of the greatest scientific-industrial projects of the war. The two Australians made science. And science, in killing and in healing, made war and then made peace, and so made history.

14

Smiling public men

Twenty-nine years later, and another goodbye. But this farewell was not happy. Florey unleashed on Oliphant as they paced up and down the Victoria Station railway platform in London, telling him he was a bloody fool. 'You know, if you leave this country you'll be committing scientific harakiri'.

Florey had come down from Oxford to see off Oliphant and his family on their journey to Australia, where Oliphant was to embark on a new chapter of life at the Australian National University (ANU). But first, Florey tried one last time to persuade his old friend to stay.

'It'll be the end of your research career!' he pleaded. 'You'll just find yourself so involved with setting up this new university that you'll have no proper time to devote to your scientific work'.

Florey believed Oliphant was foolish to leave behind his promising work in postwar Britain for a risky venture in

Canberra. There were still too many uncertainties, too many loose ends. 'You know what you'll find when you get there?' he asked, as they waited for the train to take the Oliphants to the port of Southampton. 'A lot of promises and a hole in the ground'.

Oliphant understood – and shared – many of Florey's concerns. He knew that his career might founder, his time consumed by administration, advocacy and mentorship rather than his first love: the pursuit of pure science. But there was no turning back now.

There was a fork in the road for Florey and Oliphant in 1950. While the afterglow of wartime contributions followed them both in life and work, it shone brighter for Florey. Unleashing the power of the atom was a far more ambiguous gift to humanity than releasing the healing power of penicillin, and hardly redeemed by the benefits of an all-seeing radar. Their postwar choices also made a difference.

《 • 》

It was a glittering occasion.

The King of Sweden presided over the awards ceremony on 10 December 1945, the anniversary of Alfred Nobel's death, the first one commemorated since the outbreak of war. But the atmosphere inside Stockholm's Konserthus was as chilly as the winter night outside. Discord marred what should have been a joyful celebration of a scientific triumph, collegiality reduced to formal courtesies. The three co-winners and their wives sat near each other but hardly spoke a word. This was no Peace Prize. Congratulatory handshakes were perfunctory, small talk now impossible.

Chain could not forgive Florey. He saw his behaviour as high-handed and was critical of his naivety in failing to patent penicillin. Florey, no longer close to Chain, remained aloof. In turn, both Florey and Chain were appalled by Fleming's immodesty, if not 'downright dishonesty', and refused to speak to the self-serving Scot. Fleming, aware of their resentment, kept largely to himself.

While much of Europe now lay in ruins, neutral Sweden had prospered during the war, a contrast that added poignancy to the proceedings. As champagne glasses clinked in congratulations, attending luminaries could not help but reflect on the Janus-like duality of science.

'In a time when annihilation and destruction through the inventions of man have been greater than ever before in history', remarked Professor Göran Liljestrand of the Royal Caroline Institute, 'the introduction of penicillin is a brilliant demonstration that human genius is just as well able to save life and combat disease'. Professor AHT Theorell, Director of the Department of Biochemistry at the Nobel Institute of Medicine, also highlighted the paradox of timing: 'Penicillin was made available to mankind during the biggest of wars; but it is unable to serve anything but peaceful purposes. It cannot kill a mouse, though it can heal a man'.

This was a much-needed reminder that for all the controversy over precedence and credit, penicillin itself had already saved millions of lives and loosened infection's miserable grip over millions more. More importantly, penicillin was only the beginning, the first chapter of the antibiotic revolution in public health.

Florey continued to play a part in this great endeavour. Aware of increasing bacterial resistance to penicillin as well as its ineffectiveness against some dangerous and common bacilli, he continued to search for new antibiotics. In 1948, Florey's Oxford team examined a mould found in a sewage outlet off the coast of Sardinia. After years of exacting research, they eventually managed to isolate its powerful antibacterial component, named Cephalosporin C. It spawned a new series of drugs that overcame some of the deficiencies of penicillin. By then, the Dunn School was first among many institutions around the world working to develop the next generation of antibiotics.

'You have become the most famous doctors in the whole world', Professor Theorell told the three Nobel laureates. Despite his natural reticence and lingering suspicion of the media, Florey was now a prominent figure. He never approached Fleming's public profile, but outstripped him in professional recognition. In demand for contributions as a researcher, administrator, academic leader and speaker, Florey was reluctantly transformed into what the great Irish poet William Butler Yeats called a 'smiling public man'.

《 • 》

Oliphant heard the news of Hiroshima while holidaying in Wales with his family. The next day, Tuesday, 7 August 1945, his photograph appeared on the front page of London's *Daily Telegraph* under the banner headlines: 'ALLIES INVENT ATOMIC BOMB: FIRST DROPPED ON JAPAN' and 'Men Who Made it Possible'. There was no hiding his role or responsibility anymore. Oliphant was a public figure now, and, like it or not, somewhat of a hero.

Oliphant's own uncertainty as to his status – part war hero and part war criminal – would haunt him for the rest of his life. The full extent of his involvement in the atom-bomb saga was glossed over as time went on, other names – Oppenheimer, Groves, Lawrence – becoming synonymous with the birth of the bomb. But he was never completely forgotten and he never completely forgave himself. Smiling came to him more naturally than it did to Florey, but good conscience did not allow him to cheer the fruits of his wartime advocacy and intellectual labour. He found the brilliant science and its devastating consequences impossible to reconcile.

Others were not so troubled. In April 1980, while at work on the first biography of Mark Oliphant, co-author Stewart Cockburn was searching the United States National Archives in Washington. There he found a file listing the proposed award of 96 Medals of Freedom, established by President Harry Truman a month before the atomic bomb was dropped on Hiroshima, to honour civilians whose actions aided the United States and its allies during war. All 96 on the list were 'eminent scientists of the United Kingdom, Canada, Australia and Russia', among them Otto

Frisch, Rudolf Peierls, John Cockcroft and Sir Charles Darwin, the scientific liaison officer at the British embassy who escorted Oliphant in his early meetings with Lyman Briggs and co-ordinated American, British and Canadian efforts during the Manhattan Project.

But only one Medal of Freedom was proposed with Gold Palm, the highest of the four grades of the decoration. That was to be awarded to Mark Oliphant. His citation read:

> Doctor Marcus Laurence Elwin Oliphant, Australian citizen, during the period of active hostilities in World War II, rendered exceptionally meritorious service in the field of scientific research and development. He brilliantly conceived, developed and perfected the cavity magnetron, an important factor in the entire microwave radar program of the United States, and in his further collaboration in scientific research, he made outstanding contributions in the development of the atomic bomb. Through his exceptional scientific knowledge and resourcefulness, Doctor Oliphant contributed immeasurably to the success of the Allied war effort.

Oliphant never received this high honour and only learned of it when informed by his biographers in 1980. What happened was this. The postwar Chifley Labor government had decided to stop the practice of bestowing honours on civilians. Fearing a foreign anomaly, they did not agree to those who were described as 'Australian citizens' receiving the award. Ironically, when the decision was made in 1947 there was no such thing as Australian citizenship; Oliphant was legally a 'British Subject'. Other Australians who were – accurately – so designated by the Americans, such as 'Australia's Greatest War Heroine' Nancy Wake, received their Medals of Freedom. Oliphant did not.

Unaware of just how highly regarded he was by the man who had recommended him for the medal, Major-General Groves and others in American officialdom, Oliphant returned to Birmingham in April 1945 planning to resurrect his prewar research and once again be at the forefront of nuclear science. But he never regained his scientific momentum. His precious cyclotron, plagued by construction difficulties, was only just working when Oliphant left for Australia in 1950. The synchrotron, an innovation he championed, flashed its first beam in 1953, by which time Oliphant was long gone. Even so, the Birmingham synchrotron was only the second of its kind to be built and much less expensive than the world's first in New York.

This revolutionary technology would dominate research into the structure of matter for decades to come. Oliphant's synchrotron could accelerate particles with up to 40 times more power than Lawrence's Berkeley cyclotron. It eventually inspired the world-famous Large Hadron Collider, the 27-kilometre-round giant in Switzerland. Though Oliphant first proposed a synchrotron in a classified memo to the British government in 1943, his priority as an inventor was compromised with a paper detailing the idea in a Soviet journal in 1944 and by an American publishing the concept in 1945. Nevertheless, given the importance of Oliphant's design to the subsequent development of particle accelerators and, thus, advanced atomic research, his biographers Stewart Cockburn and David Ellyard argue that Oliphant's insight might have been worthy of the Nobel prize.

《 • 》

HC 'Nugget' Coombs, one of Australia's most outstanding and influential public servants, dropped in to see Florey in Oxford in 1946 to sell him on an idea of a national university composed of four research schools. 'We wanted Florey, and we wanted Oliphant, and others of highly respected standing around whom this thing could be built', Coombs recalled later. Four eminent scholars would advise the government in shaping the new, entirely postgraduate, Australian National University in Canberra, and become directors of their respective institutes, with Florey heading a school of medical research and Oliphant a school of physical sciences. New Zealand–born anthropologist Professor Raymond Firth, and historian Keith Hancock, Florey's old friend from his early Oxford years and, until recently, a colleague of Oliphant's at Birmingham, were also approached. At first, they all said yes.

After two years of planning, over a sunny Easter weekend in 1948, the four professors, dubbed the 'maestros', visited the picturesque site of the proposed university at the foot of Black Mountain in Canberra. Photographs and colour film taken at the time show Florey and Oliphant in their influential prime. Neither yet 50, their fame and accomplishments, combined with a confidence born of holding their own with the best of Britain and the United States, lent them both presence and bearing. They could not be missed; they would not be ignored. The ANU project, with its ups and downs, was to draw them closer together, the men remaining close and in frequent correspondence for the rest of their lives.

Florey was not shy in using his hard-won eminence to achieve worthy goals. He was happy to assist in establishing the new university, believing that it would nurture Australian research and grow the country's standing in the world. He was a great believer in Australia – its people and its potential – and a fan of its idyllic lifestyle (Hancock recalled bumping into Florey on the beach at Cronulla, 'riding the breakers and browning himself on the sand, eating peaches and murmuring ecstatically, "Man *can* live by bread alone"'). Florey felt the pull of his homeland but could not break free from the attractions of his scientific career in Britain.

He gave his reasons for changing his mind. They centred on authority, independence and money – the typical trifecta for senior academics. He was never convinced that life in Canberra would suit him. His professional life had been in Britain where he was renowned in the world of scientific research and highly respected. He led a life of research work, never happier than in his laboratory, and he was well supported at the Dunn School. In Australia, it would be quite different. His eminence made him too valuable as a leader and mentor. Life in Australia would inevitably mean more administration, more bureaucratic turf wars, more distractions. Last but not least, Britain – and increasingly America – was the centre of action; Australia, in the world before emails and cheap overseas travel, at the periphery. He continued to advise the John Curtin School of Medical Research, as it was named, and officially opened it in 1958. He succeeded Mark Oliphant's friend, Sir John Cockcroft, as the ANU's third Chancellor in 1965, performing the duties remotely from Oxford.

《 • 》

Oliphant never publicly criticised Florey for choosing to remain in Oxford, but he was disappointed by his decision. Oliphant believed Florey would have been a magnet for young researchers as well as government and private funding. However, his own path to the ANU shows the limits of scientific star power in Australia at the time. 'It may be', Oliphant mused much later, 'that for science generally Florey's final decision was the best'.

In 1946, while attending the Commonwealth Prime Ministers' Conference in London, Prime Minister Ben Chifley had taken the opportunity to get to know the famous physicist. Over a walk in the park and then dinner at the Savoy Hotel, Oliphant 'was at his spell-binding best', weaving in the excitement of the Manhattan Project with a future dominated by atomic energy, where Australia could play a leading role. 'The impact on Chifley', Coombs later recalled, 'was tremendous'.

Following their meeting, Oliphant was invited to serve as a technical advisor for Australia's delegation to the United Nations Atomic Energy Commission. Although he enjoyed the role, the conference achieved little. The United States, not counting on spies in their midst, believed the Soviets were years behind in nuclear technology. Among the spies were Alan Nunn May, an undergraduate in Oliphant's Birmingham department; and Klaus Fuchs, who worked with Rudolf Peierls in Birmingham. Fuchs accompanied Peierls to New York and later Los Alamos where he enjoyed full access to vital information, making his betrayal perhaps the most devastating of all the atomic spies.

Oliphant knew both May and Fuchs, though he was not close to either. He gave Fuchs a job at Birmingham. As with Frisch and Peierls – and for the same reasons – he got him to work on the bomb rather than radar. This association, along with Oliphant's own history of outspokenness, lack of discretion and disregard for security, meant that many in American officialdom never trusted him again. He was denied a visa to travel to a nuclear physics conference in Chicago in 1951 and smeared in intelligence circles as someone whose judgment might not be sound and whose prudence could not be relied on. Oliphant was not the only person to fall victim to the Cold War. Around the same time, Florey's colleague, Ernst Chain, was also denied a visa. The State Department offered no explanation, but it was widely assumed to be because of the time Chain had spent in Czechoslovakia in the late 1940s.

Oliphant had reservations about returning to Australia. He did not want to leave Birmingham just as his cyclotron had begun to work and the synchrotron was well on the way. But he believed his homeland deserved his best: 'I said I would come because I was ... a very loyal Australian, and thought that I owed it to Australia, which had been very good to me and helped me to get this scholarship to go abroad', he recalled later.

In the end it was only Oliphant and, eventually, Hancock, who came back to head their respective research schools at the ANU. Firth, like Florey, changed his mind early on; Hancock arrived in 1957. Yet, coming home was more difficult for Oliphant than for the others. All they needed were able colleagues, a good library and, for Florey, the tools of a laboratory. Oliphant required new hardware and expensive technology to conduct advanced nuclear research. In Birmingham he left behind the cyclotron and a synchrotron under construction. He now had to start again on the other side of the world.

《 • 》

Having decided to remain in Oxford, Florey recommitted to research and public service. In 1960 he was elected President of the Royal Society, the first Australian so honoured. The president does not simply chair meetings of the society's council but is the public face of a highly respected institution that is influential with government, universities and industry. Florey may have detested flummery, but he could tolerate it to advance the causes he believed in. Sixty years ago, the presidency was the greatest honour that might be bestowed on a scientist in Britain. By all accounts he did the job superbly.

In a move that surprised his friends, Florey also accepted the role of Provost of Queen's College, Oxford. It was a great change for him, coming as those august university roles do with superintending the ungovernable, nursing the incurable and seeking consensus in the ever-disagreeable. If politics as a profession is fraught with dishonesty and disloyalty, nothing quite matches the deviousness of college politics. Somehow, Florey managed.

Lab work increasingly took a back seat to administrative duties, confounding the initial expectations there would be less of them at Oxford than at ANU. Yet the honours continued to roll, even as his scientific output slowed. Upon nearing the conclusion of his term as President of the Royal Society, he was created a Life Peer as Baron Florey of Adelaide and Marston. 'It is good to know that some real science is to be infiltrated into the House of Lords', Oliphant congratulated him. The honour neatly recognised the two locations for which he harboured the most affection: the place he grew up and his home near Oxford. Not yet done, the queen also conferred on Florey the rare and personal gift of the monarch, the Order of Merit, signifying his particular eminence.

He appreciated the recognition but did not crave it. In a letter to Mellanby, he wrote, 'I could quite readily dispense with all the publicity and "honours" and such-like, in fact I would readily surrender it all for a little sustained peace of mind such as we had when working … before the war'. But he could not turn back the clock. Fame increasingly sidetracked his efforts, denying him the greater pleasures of scientific pursuit and the world, perhaps, the benefit of more discoveries. Florey, a great technician and team builder, had no problem attracting collaborators and students to solve scientific problems. His colleague and biographer, Gwyn Macfarlane, concluded that 'he was the best and soundest builder of solid factual knowledge in British experimental pathology'. He might not have been the easiest to work with, but he was among the most sought after.

Ethel Florey's health had been poor for many years. Increasing deafness and heart disease slowed her down, but she continued to travel and lecture on penicillin and her work. Despite a heart attack and warnings from both her husband and her doctors, she would not lighten her load. She died in October 1966, not long after returning from the wedding of her son Charles in the United States.

Oliphant wrote a letter of warm sympathy to Florey. On 29 October, Florey replied. It was among their last correspondence, with Florey mustering as much intimacy as his natural formality would allow, leavened by over half a century of friendship:

My dear Oliphant,
Thank you for sending such a kind and sympathetic message. As you say when death comes it is a shock whatever the circumstances but it is pleasant to have messages from friends.

Thank you so much for writing,
Yours,
Howard Florey

Florey, tired and grieving, found personal happiness late in life. He married his Dunn School colleague Margaret Jennings, in June 1967. But his own health suffered too. He had been diagnosed with angina in 1950; his doctor, unable to do anything for the man who had done so much for so many, advised him to rest. This was easier said than done for someone of Florey's calling and temperament. He was never again fully well, but the former youth athlete and competitive sportsman soldiered on. For those last few months, he was genuinely happy. His heart finally gave out on 21 February 1968. Florey was 69. He was outlived by millions around the world, alive thanks to his gift of magic mould.

« • »

When Oliphant arrived to see the new laboratories at the ANU, he found a big hole in the ground, only partly redeemed by the appearance of foundations. It was exactly as Florey had predicted.

But it was too late to back out. Oliphant became the first Director of the Research School of Physical Sciences. He recruited talented colleagues, fought the university and government bureaucracies, lobbied politicians and established a leading centre of physics research. But his own research was compromised. The time spent as a leader, administrator, construction supervisor, recruiter, advocate and stirrer left less and less time for his own scientific work (yet still he continued to add to his duties, helping found and becoming the first President of the Australian Academy of Science). He could never recapture those golden days of tinkering with apparatus at the Cavendish under Rutherford or the exhilaration of the radar work at Birmingham or the collaboration with Lawrence on the bomb at Berkeley. Oliphant wistfully recalled the moment he ignored Florey's pleas at Victoria Station, 'I carried on, and I did ruin my research career'. Both Florey and Oliphant were victims of their own success: what made them revolutionary figures – their talent as team leaders and lobbyists – now came to almost completely overshadow their first passion, the work of scientific discovery.

Oliphant's relations with the university and the government quickly deteriorated as it became clear he would not get the necessary money to build an internationally competitive particle accelerator. He struggled for years to complete a powerful proton-synchrotron all the time falling further behind similar projects in Europe, the United States and the Soviet Union. Competitors cruelly dubbed Oliphant's machine-building project as 'the White Oliphant'. A serious accident to a staff member working on equipment in July 1962 added to his woes. His resignation as director a year later came as no surprise. Burned out by the labours of administration and the fires of combat, he had finally had enough.

Later, when asked whether he regretted moving from Britain to the ANU, Oliphant replied, 'For myself I regret it. I realise each year when I go back to Cambridge just what I gave up in leaving it'. He did not formally forfeit his position as professor until reaching retirement age, whereupon the university conferred upon him the title of Professor Emeritus and a small office. Other pursuits now filled the void. As a fluent and enthusiastic public speaker, he received invitations from around the world to lecture and broadcast. Outspoken, sometimes controversial, always engaging, he was in demand as a scientist and increasingly as a commentator. While many are brought up by wary parents to never discuss religion and politics, they were staples in Oliphant's repertoire: he was not afraid of discussing God or the clergy, or political issues. Discretion or reticence was never Oliphant's strong point.

He continued to struggle with the legacy of his wartime work. Oliphant described himself as a 'belligerent pacifist'. In 1945 he very likely disagreed with dropping the bombs on Hiroshima and Nagasaki, but once the atomic genie was out of the bottle, he did not believe America should enjoy a nuclear monopoly. In the first postwar decade he advocated for an independent nuclear capability for Britain *and* Australia. Later he soured on nuclear weapons altogether and joined with the Pugwash Conferences on Science and World Affairs, attending the first conference of this international association of scientists dedicated to nuclear disarmament, in Nova Scotia, Canada, in July 1957. Poignantly, this was close to the location of his forced landing in August 1941, when he travelled to convince America that the atom bomb should be built. The postwar consensus defined nuclear scientists, in the words of Robert Oppenheimer, as destroyers of worlds, a far cry from their self-regard as the poets of modern science. Pugwash and similar initiatives gave Oliphant, and many others who felt like him, an opportunity to atone for their wartime activities and help ensure atomic fission would never kill again. For a time Oliphant continued to support the peaceful use of nuclear energy, but he eventually become disillusioned with that too. Late in life he became a convert to renewable energy.

After it had become 'quite clear to me that the university … would rather have the space that I occupied than my company', Oliphant vacated his tiny office and the Professor Emeritus duties and accepted an appointment by South Australia's new premier, Don Dunstan, as the state's next governor. Dunstan, an unconventional figure himself, was not warned off by Oliphant's admission that his 'approach may differ considerably from the accepted norm'. Despite Rosa's early reservations, Oliphant enjoyed his five years as governor. Refusing to function in a purely ceremonial capacity, Oliphant regularly spoke out on issues ranging from the dangers of nuclear testing to environmental protection and the tension between science and religion. And he did not hesitate to express anti-libertarian and anti-republican views, much to the chagrin of the progressive premier. His outspoken behaviour and disregard for protocol were at times highly controversial but also won him many admirers.

His public life and advocacy of numerous scientific and political causes close to his heart continued for decades after the end of his gubernatorial term. He was, says his grandson, 'a lifelong activist'. As the 1980s turned into the 1990s, Oliphant had the satisfaction to witness the waning of the Cold War, and if not quite the end of the nuclear nightmare, then at least major efforts towards nuclear disarmament.

Oliphant's personal life was by and large happy, if punctured by occasional tragedy. His 3-year-old son, Geoffrey, had died of meningitis when he was at the Cavendish, and another son, Michael, died aged just 35, in January 1971, after years of poor health. His beloved wife Rosa passed away in 1987, after many years during which Oliphant cared for her as she struggled with dementia. Rosa's fight caused Oliphant to campaign for voluntary euthanasia in his final years.

On 14 July 2000, Oliphant died in Canberra. He was 98-years-old, his life spanning the entirety of the turbulent 20th century he did so much to shape.

Afterword

If you cross Oxford's High Street as you leave Florey's Queen's College and walk south-west for about a kilometre you will arrive at the Botanic Gardens, the oldest in Britain. There, just opposite Magdalen College, stands a small stone memorial, an inscribed slab buffeted by a sea of flowers. It reads:

> This rose garden was given in honour of the research workers in this university who discovered the clinical importance of penicillin.
>
> For saving of life, relief of suffering, and inspiration to further research, all mankind is in their debt.
>
> Those who did this work were: EP Abraham E Chain CM Fletcher HW Florey ME Florey AD Gardner NJ Heatley MA Jennings J Orr-Ewing AG Sanders.

Afterword

'One can think', wrote Australian immunologist and Nobel laureate Sir Macfarlane Burnet, 'of the middle of the twentieth century as the end of one of the most important social revolutions in history – the virtual elimination of infectious diseases as a significant factor in social life'. Standing at the pinnacle of that revolutionary moment, Howard Florey is unquestionably Australia's greatest benefactor to humanity. No other Australian has had a greater impact on the welfare of the world.

Today, we take antibiotics for granted. This makes it difficult to fully appreciate the stunning impact on human health and healing that flowed from that first experiment with mice in May 1940. One estimate puts the number of lives saved by penicillin since 1944 at 200 million. This means penicillin saved not only more lives than were lost in the Second World War but twice as many lives as were lost to war and political violence in the 20th century.

The count for penicillin's antibiotic descendants is truly staggering. And, despite growing concerns about antibiotic resistance, it continues to climb. In the first 20 years of the 21st century, an estimated 66 million men, women and children suffering from tuberculosis alone owed their lives to antibiotics. Antibiotics account for around 23 years added to our life expectancy in the last century. The reduction in suffering and the improvement in quality of life that antibiotics have brought to billions of people around the world is incalculable. And yet, Florey never set out to save humanity from pain and misery. As a scientist, he was always more savant than saviour, his endeavour charged by scientific curiosity and only later leavened by hope and possibility.

Nestled in the Oral History Collection at the National Library of Australia is one of the few recordings of Florey's voice, an interview conducted in Canberra on 5 April 1967. It was his last visit to his homeland, when he was, he told friends, 'living on borrowed time'. In less than 12 months he was dead. His reflections, alive with modesty yet firmness, capture the paradox of Florey the man and the scientist: a loner who promoted research teams; a maven with few peers but no appetite for pomposity; an avid contributor to human knowledge but an accidental benefactor to the human condition:

> People sometimes think that I and the others worked on penicillin because we were interested in suffering humanity. I don't think it ever crossed our minds about suffering humanity. This was an interesting scientific exercise, and because it was of some use in medicine is very gratifying, but this was not the reason that we started working on it.

Lennard Bickel, his first biographer, wrote:

> Florey was totally averse and reluctant to wear a mantle of glory. He would have protested at the suggestion that his life's work reached more people than did the teachings of Jesus Christ. Yet, by the time

Afterword

he died, there was hardly a family on the face of the planet which had not been touched by what he did, and by what he set in motion.

His success can be seen not in statues or monuments – though there are a few of those, not least a memorial stone at Westminster Abbey, next to the astronomers John and William Herschel and the naturalist Charles Darwin – but in the eyes of the Australian soldier wounded on a foreign battlefield, the young mother in Senegal recovering after a difficult childbirth, the baby sick with meningitis in Peru, the grandfather struggling with pneumonia in Turkey, and the teenage boy burning with infection in Laos. The gift of penicillin touches us all, though almost none of the beneficiaries will ever know the name of the reticent Australian who made it all possible.

Howard Florey did not simply help save hundreds of millions of lives, he changed the way we see our world. He gave humanity hope. From childbirth to old age, our world was no longer so threatening or our existence as vulnerable. Florey allowed us to believe – for the first time with good reason – that a world without sickness might one day be possible; that the empire of disease might yet be conquered by human ingenuity. After Florey, the world was no longer inevitably a hospital or an infant cemetery. It was, as he said, a miracle.

《 • 》

Mark Oliphant was born in the first year of the 20th century and died in its last. The bloodiest century in human history was also the century of the greatest human progress. As humankind peered into space and stared into microscopes – both the infinite and the infinitesimal coming into view and revealing unbelievable marvels – it also found time to launch technologies that unleashed unimagined carnage. Oliphant's life is the story of this paradox.

Microwave radar, powered by a resonant cavity magnetron from Oliphant's laboratory, was the single most important technological innovation of the war. Even Richard Rhodes, celebrated author of two books on the making of the atomic bomb and the hydrogen bomb, concedes, 'The atomic bomb was a sideshow in World War II compared to radar'. Microwave radar was the 'one scientific weapon [that] meant the difference between victory and defeat' – it swung the balance against the Axis powers and helped to roll back the aggressors all the way to their ultimate defeat. Through four years of hard and bitter fighting in the air and at sea, the technology born of Oliphant's lab gave the Allies an incomparable edge, a new set of eyes that could locate and pinpoint enemy aircraft, naval vessels, submarines, as well as ground targets, at any time and in any conditions – and then destroy them, often using proximity fuze–guided munitions that did not miss their mark. The enemy, by comparison, was largely fighting blind.

Radar's utility did not stop when the war ended. It is difficult to imagine our modern world of mass air travel

Afterword

and mass transport without it. The cavity magnetron helped kick-start the postwar age of globalisation by making large-scale, coordinated movement of people and goods safer than it has ever been. 'Pulses into Ploughshares', as Watson-Watt quipped. The number of lives and property saved by radar is difficult to estimate, but not insubstantial. RMS *Titanic* would have safely reached New York on its maiden voyage, with the radar able to 'see' icebergs a hundred miles away. The pioneering era of air traffic control, sea shipping navigation, aircraft anticollision, space surveillance and rendezvous systems relied on technology descended from the first Birmingham magnetron. As did novel uses developed after the war in astronomy, meteorology and even geology, with ground-penetrating radar for scientific observation and mineral exploration. From the furthest reaches of our solar system to the hidden depths below the Earth's surface, radar has helped scientists see further and clearer than Spitfire and B-29 pilots could have ever dreamed.

Radar has penetrated our homes too. Percy Spencer, one of the top American engineers working on cavity magnetrons for Raytheon during the war, noticed one day that a chocolate bar in his pocket melted while he was standing next to a working radar set. Intrigued, he investigated other food – including popcorn – and the rest is history. First sold as 'Radarange' in 1946, one billion microwave ovens are used every day around the world.

The impact and legacy of Oliphant's second wartime contribution has been far more controversial. The atomic bomb helped to end the war in 1945 instead of 1946, causing great devastation but also saving millions of casualties.

Long before the war, nuclear physicists who were searching for the fundamental building blocks of the universe knew that the atom housed energy that might be unlocked in the future. In early 1940, Oliphant realised the future was now. The Fisch–Peierls memorandum convinced him that nuclear fission made nuclear weapons feasible and therefore probable. Once he knew the super-weapon was within reach, he relentlessly drove the quest to unlock and unleash the atom's energy against Hitler before Hitler could unleash it against everyone else. Oliphant was the critical spark that forced Britain and the United States to confront this new nuclear reality. 'Oliphant's heroic efforts are generally felt to be the "catalyst" that finally pushed the American bomb effort over the top', concludes the US-based Atomic Heritage Foundation. Many later claimed to have inspired the development of the atomic bomb. But Oliphant first dived into the scientific, bureaucratic and political swamps. He found out how deep the water was. Only then did others follow.

After 8:15 am on 6 August 1945, the world was a different place. Japan suffered a catastrophe that continues to haunt us all. The nature of war and politics and international relations – perhaps even human consciousness itself – changed forever. Not since the religious fervour of the Middle Ages had apocalypse seemed so possible. For all the developments of science and the bumpy progress of democracy, the postwar world was dominated by the knowledge that humankind could now destroy itself. The American writer William Faulkner put it best in his 1950 Nobel Banquet speech, saying simply, 'There is only one question: When will I be blown up?'

Even Winston Churchill, an agnostic and generally an optimist, saw 'nothing so menacing to our civilization since the Mongols'. In the shadow of nuclear weapons, he found it 'poignant to look at youth in all its activity and ardour and, most of all, to watch little children playing their merry games, and wonder what would lie before them if God wearied of mankind'. Many did indeed wonder and worry, Stanley Kramer's *On The Beach*, based on Neville Shute's post-apocalyptic book set in doomed Melbourne, and Stanley Kubrick's *Dr Strangelove* testifying to the popular fascination and dread of the nuclear threat.

Yet as generations were born, played, worked, loved and lived in the shadow of a mushroom cloud, some argued that the threat posed by nuclear weapons kept the Cold War from turning into a hot one. Many believe that the prospect of such 'Mutually Assured Destruction' had a chilling effect on hot political tempers. In the view of John Lewis Gaddis, the preeminent American historian of the Cold War, 'It seems inescapable that what has really made the difference in inducing this unaccustomed caution' between the Soviet Union and the United States 'has been the working of the nuclear deterrent'. As with everything else to do with the atom's hidden power, this has been a contentious view.

Nuclear warheads, of course, were not the only offspring of the Frisch–Peierls memorandum. As Briggs' Uranium Committee initially contemplated, nuclear fission had peaceful uses – boilers, not bombs. Nuclear energy, at its peak in 1996, provided almost 18 per cent of global electricity. But 'atoms for peace' have always been controversial, partly because of the inevitable association with nuclear weapons and partly out of public safety and environmental concerns, amplified by a number of high-profile accidents like Three Mile Island, Chernobyl and Fukushima. While its popularity is on the decline, research and development work continues to make nuclear reactors safer, smaller, cheaper and more efficient. Proponents, including environmentalist Michael Shellenberger and nuclear historian Richard Rhodes, argue its emissions-free nature makes nuclear energy the only currently viable alternative to fossil fuels as the world confronts the challenge of climate change. Not everyone is convinced. Mark Oliphant, for one, came to prefer renewable energy as the solution.

Oliphant was proud of his wartime work as a triumphant and transformative scientific achievement, but haunted by his role in developing the bomb. Nuclear weapons forced us to confront our own demise. Civilisation and the survival of our species – as well as all others – are now subject to our passions. This responsibility has rested uneasily with humankind for 75 years. Oliphant, who helped to create this reality, fought against it with guilt and rage for the rest of his life. In retirement he reflected: 'I still think …

that science can save mankind, but only with the willing cooperation of man'.

《 • 》

As different as their respective scientific contributions to the world have been, Florey and Oliphant were both the harbingers and avatars of Big Science. 'Florey', said Dame Sally Davies, formerly the Chief Medical Officer for England, 'knew not only his science, and how to build this multi-disciplinary team that was well before his time, but how to go out and get the money and make it happen'. The same could be said of Oliphant, a scientist of substance, to be sure, but also a good team builder and leader, an evangelist, lobbyist and advocate.

Florey and Oliphant did not invent a new system of doing science – arguably this was pioneered by German scientists working at the intersection of research labs and industrial conglomerates in the early years of the 20th century – but after them nothing would be the same. Gone were the days of solitary and secluded savants braving the unknown like solo sailors, their 'string and sealing wax' operations running on sleepless nights and the smell of an oily rag. Big Science meant big teams, big budgets and big expectations. The development of penicillin, microwave radar and the atomic bomb were the first multi-billion-dollar scientific projects. Carried on a mass scale across vast networks of institutions, blurring the boundaries of research and manufacture and transcending national borders, the 'big three' presaged and ushered in the new postwar era of scientific progress, from drug development through space exploration to the information revolution.

For two men who spent their early careers having to beg, borrow and steal a few hundred pounds in funding, it must have been fascinating, if not entirely gratifying, to witness the transformation. While passion and a searching mind advanced knowledge, the labour of research increasingly focused on outcomes and their monetisation rather than the intellectual quest. The Oxford-to-Peoria and the Birmingham-to-Manhattan sagas provided the blueprint for the new triumvirate of science, government and business. This too is part of Florey and Oliphant's legacy.

Their scientific work in Britain during a hundred days in 1940 and their advocacy in the United States during a second hundred days in 1941 ensures their place in history. Both men changed the way we see ourselves. After Florey, the world was a little less threatening and human beings a little less vulnerable. After Oliphant, the world was more threatening and human beings more vulnerable. The two friends from Adelaide helped to win the most destructive war in history, and their work continues to shape our lives today.

《 • 》

Howard Florey arrived in Oxford 100 years ago, in the northern winter of 1922. Five years later, Mark Oliphant first knocked on the doors of the Cavendish Laboratory in Cambridge. Within 20 years they would become the two

Afterword

most consequential Australians of the Second World War. The legacy lives on in the better health and greater wealth of nations. Not bad for two boys from Adelaide.

But Florey and Oliphant's contribution goes beyond world war and peace, to matters closer to home. Their achievements helped change the way we think of ourselves, see our prospects and view our future. They helped Australia to find a place in the sun; not a lucky country, but a smart and self-assured one.

To Florey and Oliphant, like many men and women before and since, the decision to venture overseas in pursuit of their goals came naturally. Australia was lacking in scientific and intellectual standing – not to mention necessary resources and infrastructure – as well as social and cultural confidence. Tyrannised by distance and a narrowness of vision, many Australians saw their country as an intellectual backwater.

In his 1958 essay 'The Prodigal Son', Patrick White wrote of the 'Great Australian Emptiness, in which the mind is the least of possessions … in which beautiful youths and girls stare at life through blind blue eyes'. Not only, our Nobel laureate for Literature argued, was Australia infertile to the life of the mind but represented a great failure of imaginative nerve. Australians, White believed, were passive outsiders at the margins of history.

The yearning for more, accompanied by a conviction that life at home was somehow less meaningful than elsewhere, bred a sense of irrelevance. The Melbourne-based critic AA Phillips diagnosed it as a 'disease of the Australian mind' and dubbed it 'the cultural cringe'. Escaping a 'stifling intellectual torpor', as actor Barry Humphries described it, had by the 1960s become a rite of passage for another generation of intellectual exiles. They found the Europeans more worldly and the Americans more worthy. Recalling an observation by Yeats that for his Irish compatriots England was fairyland, Peter Craven writes that 'for an Australian, everywhere was fairyland'.

It was not just jaded radicals and disaffected bohemians who suffered frustration. Even the decidedly buttoned-down and conservative Howard Florey recognised that 'to the outsider ... Australia is a very conservative country ... new ideas are not common here'. If history was not – could not be – made in Australia, to make history Australians had to leave and be someone else, somewhere else. 'Can't we do anything ourselves as Australians?', Manning Clark was still asking in 1987, in the preface to the sixth and final volume of his *History of Australia*.

This is no longer the case. Australia is a success story. We have been more successful economically, politically and socially over the last 35 years and have become a more self-assured nation, 'much more happy in our skin', says novelist David Malouf. Australians, particularly young Australians, are confident about their country and themselves. 'Beautiful youths and girls' are no longer blind, their stares no longer vacant. As a nation, we relish the opportunity to engage, contribute and lead.

Afterword

Over the past century, 13 Australians have been awarded a Nobel prize, a majority, coincidentally, in Medicine or Physiology. Our first three laureates – the father–son physicist team of William Lawrence Bragg and William Henry Bragg, and Florey himself – carried out their Nobel-winning work overseas. But only three (scientists John Cornforth and Elizabeth Blackburn and writer JM Coetzee) of the rest did so. They include familiar names like Peter Doherty and Barry Marshall. Young, talented, and ambitious Australians no longer need to leave our shores in order to pursue a life of intellectual and scientific excellence. Some Australian universities and research institutions rival those elsewhere and Australia is a global leader in medical research. Australians can now change history without leaving home. Two of our Nobel laureates, Brian Schmidt and JM Coetzee, were born overseas and have chosen to call Australia home. Florey and Oliphant could only have dreamed that Australia might one day attract the best and the brightest from around the world.

Florey and Oliphant had the courage of their hopes and convictions. They had dreams bigger than their country could nurture. Their lives were enlarged by high purpose, and they were never happier than when chasing those dreams. They pursued their purpose with grit, overcoming obstacles and disappointments. They believed, battled and persevered, and the world is a better place for it. Australia might learn as much from their spirit as from their achievements.

Acknowledgments

One of the lessons of history, certainly in the story of Howard Florey and Mark Oliphant, is that almost nothing of consequence is ever accomplished alone. It was never truer than with this book.

Dr Arthur Chrenkoff brought order to chaos and charted a way forward. But more than that, he brought fun and friendship to a task that can be lonely and difficult. For him, my heartfelt gratitude.

This book was born before the global pandemic but was written during it. Before the shutters came down, I received generous assistance from members of The Queen's College, Oxford, of which Lord Florey was Provost (1962–1968). Former Provost, Professor Paul Madden FRS, FRSE, Elaine Evers and Professor Chris Norbury (also of the Sir William Dunn School of Pathology) guided my first steps and showed me Florey's last. Colleagues of Florey's, the late Dr Bill Frankland MBE, Dr Godfrey Fowler OBE and the late Dr Peter Neumann OBE shared invaluable insights into Florey's mature character and temperament. My thanks also to Professor Henry Woudhuysen FSA, FBA, Rector of Lincoln College, Oxford, who enthusiastically showed me college artifacts relating to Florey's time there as a Fellow. In London, Keith Moore, Head of Library and Information Services at the Royal Society, pointed me in the right direction.

Then came the pandemic. While the virus hindered primary research overseas, it could but little daunt the Digital Age. Another Australian contribution to the modern world, wi-fi, and the internet helped bridge the tyranny of distance.

In Australia we are blessed with our National Library. The doors, and now the website, of the library are the portals to our national story, including this tale of Oliphant and Florey. The library's digital collection, Trove, provided plenty of treasures. To the Director General, Dr Marie-Louise Ayers FAHA, the Council and staff of this great institution, my sincere thanks, both in my capacity as Chair of its Council and as an author.

While subsequent trips to Britain proved impossible, the Australian Joint Copying Project enabled access to materials in British archives – in particular, the Royal Society. Again, I thank the National Library's AJCP Project Manager, Emma Jolley, and her team for their generous assistance.

The University of Adelaide's library, specifically its Rare Books and Manuscripts collection, is the official depository of Mark Oliphant's papers. It is indispensable. The library also holds some of Howard Florey's, including a few of the arresting photographs in this volume. Thanks to Marie Larsen and Lee Hayes for helping me to navigate these collections.

Closer to home, the view from the State Library of Queensland brings joy to even a tough assignment. Thanks to the Queensland State Librarian, Vicki McDonald AM, and her team for providing Brisbane's most agreeable research environment.

A posse of learned professors, Peter Coaldrake AO, Don Markwell, Ian O'Donnell MRIA and John Scott provided sound advice regarding aspects of this project. I took most of it.

Anna Young, Curator (Objects), University of Birmingham, kindly assisted with materials from Oliphant's time at the School of Physics.

Thanks to Dr David Ellyard, co-author of the definitive biography of Mark Oliphant, who patiently answered my questions.

Monica Oliphant AO generously shared memories and a beautiful portrait of a young laboratory-bound father-in-law. Michael Wilson ably set the scene for his grandfather's life and achievements. 'He remained an activist', Michael reminded me.

Thanks also to Professor Charles Florey, who shared some of his father's home movies and generously spent time searching his collection for photos of his father with Oliphant.

Diana Ritch, daughter of oral-history pioneer Hazel de Berg, kindly provided her mother's unpublished and

Acknowledgments

illuminating notes made following her interviews with both Florey and Oliphant.

Along the way, Jack Fisher and Brenden Kocsis provided generous advice, while Tom and a team of master baristas and staff at Sippy Tom fuelled the flame for two busy years.

And, finally, my appreciation to Elspeth Menzies, Executive Publisher at NewSouth Publishing, for taking on a more popular and freewheeling narrative than commonly accommodated within a university press. Who says publishers are not brave? Victoria Chance edited my exuberant draft with skill. Sophia Oravecz and Joumana Awad hustled the project to a successful conclusion. Thanks also to the design and marketing teams, as well as PR guru Jackie Evans, who helped Howard Florey and Mark Oliphant's inspirational story reach their fellow Australians.

Sources

Prologue

3   'was probably the moment': AJP Taylor, *The Second World War: An Illustrated History* [1975] in *The Second World War and its Aftermath*, Folio Society, London, 1998, p. 67.

3   Florey and his team had rubbed the spores: Eric Lax, *The Mold in Dr. Florey's Coat: The Story of the Penicillin Miracle*, Henry Holt & Company, New York, 2004, pp. 125–6.

5   'I hope it never happens': Robin Hughes, *Australian Lives: Stories of Twentieth Century Australians*, Angus & Robertson, Sydney, 1996, p. 129.

6   'vast mechanized Iliad of suffering': AN Wilson, *After the Victorians: 1901–1953*, Hutchinson, London, 2005, p. 401.

Chapter 1

9   It was a farewell, of sorts: Lennard Bickel, *Rise up to Life: A Biography of Howard Walter Florey Who Made Penicillin and Gave It to the World*, Angus & Robertson, London, 1972, p. 263.

11   'the Imperial federation': *Argus*, 2 January 1901.

11   'I thought the clapping would never end': Alexandra Hasluck (ed.), *Audrey Tennyson's Vice Regal Days: The Australian Letters of Audrey Lady Tennyson to Her Mother Zacintha Boyle, 1899–1903*, National Library of Australia, Canberra, 1978, p. 133.

12   'About 64 roads to the other world': as cited in Coleman O Parsons, 'Mark Twain in Adelaide, South Australia', *Mark Twain Journal*, vol. 21, no. 3, 1983, pp. 51–4.

12   'a general air of comfort and well-being': Michael Cannon, *The Roaring Days*, Today's Australia Publishing Company, Mornington, 1998, p. 616.

Sources

12 'perfumed ... with saddle-oil': James Morris, *Pax Britannica: The Climax of Empire*, Faber & Faber, London, 1968, p. 10.
14 'in the hands of the physician': 'The Nobel Prize in Physiology or Medicine 1901', <www.nobelprize.org/prizes/medicine/1901/summary/>, accessed 10 May 2022.
15 'analysis of crystal structure': 'The Nobel Prize in Physics 1915', <www.nobelprize.org/prizes/physics/1915/summary/>, accessed 10 May 2022.
15 There were limits to local achievement: RW Home, 'The problem of intellectual isolation in scientific life: WHT Bragg and the Australian scientific community, 1886–1909', *Historical Records of Australian Science*, vol. 6, no. 1, 1984, pp. 19–30.
17 'I lived in the same suburb': Clarence Larson, *Pioneers of Science and Technology: Interview with Mark Oliphant*, Washington DC, 11 March 1988, <www.youtube.com/watch?v=z0QaXvLsR9A>, accessed 10 November 2021.
17 'Oh, you'd like to be a sort of Pasteur?': as cited in Bickel, *Rise up to Life*, p. xix.
18 'nearly blew Mitcham off the map': as cited in Stewart Cockburn & David Ellyard, *Oliphant: The Life and Times of Sir Mark Oliphant*, Axiom Books, Adelaide, 1981, p. 17.
18 'do-gooders': Robyn Williams, 'Sir Mark Oliphant', *Life and Times*, 1985, radio program, ABC Radio National, <www.abc.net.au/rn/legacy/programs/lifeandtimes/stories/2009/2588019.htm>, accessed 11 November 2021.
18 'fat prelates in Rome': as cited in Cockburn & Ellyard, *Oliphant*, p. 272.
19 Joseph Florey supplying boots for the army: Charles Florey, personal communication with the author, 27 September 2021.
20 'fondness of and success in manly outdoor sports': 'The Selection of the Scholars' in the *Last Will and Testament of Cecil J Rhodes*, 1902, <en.wikisource.org/wiki/Page:Last_Will_and_Testament_of_Cecil_Rhodes.djvu/50>, accessed 11 November 2021.
21 'on the Rhode to Magdalene': 'Last Year's Graduates', *Adelaide Medical Students Society Review*, vol. XIII, no. 21, July 1922, p. 50.
22 'dog's body': Charles Weiner, 'Oral histories: Mark Oliphant', Niels Bohr Library & Archives, American Institute of Physics, 3 November 1971, <www.aip.org/history-programs/niels-bohr-library/oral-histories/4805>, accessed 21 November 2021.
22 'laboratories smelt of tar and resin': WGK Duncan & Roger Leonard, *The University of Adelaide, 1874–1974*, Rigby, Adelaide, 1973, p. 60.
23 'I was a nobody': as cited in Cockburn & Ellyard, *Oliphant*, p. 29.
23 'I grew really to love': Robin Hughes, *Australian Lives: Stories of Twentieth Century Australians*, Angus & Robertson, Sydney, 1996, p. 119.

Chapter 2

25 'the greatest collective work of science': Jacob Bronowski, *The Ascent of Man*, British Broadcasting Corporation, London, 1973, pp. 330, 349.

25 'walking the path of God': as cited in Brian VanDeMark, *Pandora's Keepers: Nine Men and the Atomic Bomb*, Back Bay Books, New York, 2005, p. 27.

26 William Osler observation on improving human health: Henry F Dowling, *Fighting Infection: Conquests of the Twentieth Century*, Harvard University Press, Cambridge Mass., 1977, p. 10.

27 'undoubtedly the most beautiful college': as cited in Gwyn Macfarlane, *Howard Florey: The Making of a Great Scientist*, Oxford University Press, Oxford, 1979, p. 61.

27 Oxford emerging from the Middle Ages: Macfarlane, *Howard Florey*, p. 59.

27 nearly 20 per cent of students perished: John Taylor, 'How World War I changed British universities forever', *The Conversation*, 9 November 2018, <theconversation.com/how-world-war-i-changed-british-universities-forever-106104>, accessed 11 November 2021.

28 'Oxford, in those days': Evelyn Waugh, *Brideshead Revisited*, Penguin Classics, London, 2000, p. 17.

28 'Englishmen ... a queer lot': as cited in Macfarlane, *Howard Florey*, pp. 61–2.

28 'annoy me excessively': as cited in Macfarlane, *Howard Florey*, p. 68.

30 'The atmosphere is so different here': as cited in Lennard Bickel, *Rise up to Life: A Biography of Howard Walter Florey Who Made Penicillin and Gave It to the World*, Angus & Robertson, London, 1972, p. 18.

30 'damnably lonely': as cited in Macfarlane, *Howard Florey*, p. 80.

30 'I can see myself': as cited in Macfarlane, *Howard Florey*, p. 80.

31 'I couldn't possibly get a better launching': as cited in: Eric Lax, *The Mold in Dr. Florey's Coat: The Story of the Penicillin Miracle*, Henry Holt & Company, New York, 2004, p. 38.

31 'about the most brilliant scientific collection': as cited in Macfarlane, *Howard Florey*, p. 81.

32 Runciman 'with a parakeet'; Beaton 'an evening jacket, red shoes': John Costello, *The Masks of Treachery*, William Morrow & Company, Inc., New York, 1988, p. 121.

32 'In Cambridge ... there was so much to distract you': Christopher Isherwood, *Lions and Shadows: An Education in the Twenties*, Methuen, London, 1985, p. 40.

33 'mathematical gloom': as cited in Costello, *The Masks of Treachery*, p. 118.

33 'His drive and ambition were manifest'; 'A great fire seemed to burn within him'; 'He displayed utter integrity'; 'To cope with him at times': as cited in Bickel, *Rise up to Life*, p. 24.

Sources

34 'I could always get away with being audacious': as cited in Bickel, *Rise up to Life*, p. 24.
34 Florey sometimes referred to himself as 'the bushranger': For one example, though later in life, see Edward Abraham, 'Howard Walter Florey, Baron Florey of Adelaide and Marston, 1898–1968', *Biographical Memoirs of Fellows of the Royal Society*, 1 November 1971, vol. 17, pp. 255–302 at p. 283.
34 'rough colonial genius': as cited in Bickel, *Rise up to Life*, p. 27.
35 'false image' of Ethel: as cited in Lax, *The Mold in Dr. Florey's Coat*, p. 42.
35 'his work … became a barrier': Macfarlane, *Howard Florey*, p. 156.
36 'the colonial contingent': Michael Wilson, interview with the author, June 2021.
37 'The Germans state openly': as cited in Macfarlane, *Howard Florey*, p. 75.
37 'Germans, Austrians, Italians, Czechs, Hungarians': as cited in Macfarlane, *Howard Florey*, p. 78.
37 Cavendish as target of Soviet scientific espionage: Costello, *The Masks of Treachery*, pp. 115–6.
37 'Nearly two-thirds of prominent British Communists': Andrew Boyle, *The Climate of Treason: Five Who Spied for Russia*, Hutchinson & Co Ltd, London, 1979, p. 36.
40 'by far, the greatest physical laboratory in the world': Mark Oliphant, *Rutherford – Recollections of the Cambridge Days*, Elsevier, London, 1972, p. 18.
40 'decrepit old building': Robyn Williams, 'Sir Mark Oliphant', *Life and Times*, 1985, radio program, ABC Radio National, <www.abc.net.au/rn/legacy/programs/lifeandtimes/stories/2009/2588019.htm>, accessed 11 November 2021.
40 'string and sealing wax': Mark Oliphant, 'Looking back: Sir Mark Oliphant in conversation with David Ellyard', in Macfarlane Burnet & Mark Oliphant, *Sir Frank Macfarlane Burnet on Ageing & Looking Back by Sir Mark Oliphant*, Australian Broadcasting Commission, Sydney, 1979, pp. 31–2.
40 'We haven't the money so we have to think': as cited in Stewart Cockburn & David Ellyard, *Oliphant: The Life and Times of Sir Mark Oliphant*, Axiom Books, Adelaide, 1981, p. 31.
40 'a watery sun … scarcely penetrated into the room': Charles Weiner, 'Oral histories: Mark Oliphant', Niels Bohr Library & Archives, American Institute of Physics, 3 November 1971, <www.aip.org/history-programs/niels-bohr-library/oral-histories/4805>, accessed 21 November 2021.
40 'littered with books and papers': Oliphant, *Rutherford*, p. 19.
41 'spluttered a little as he talked': Oliphant, *Rutherford*, p. 19.

41 'I feel it in my water!': as cited in Robin Hughes, *Australian Lives: Stories of Twentieth Century Australians*, Angus & Robertson, Sydney, 1996, p. 123.

41 'It was ... absolute heaven to be working with him': as cited in Hughes, *Australian Lives*, p. 122.

41 six Nobel prizes traceable to 1932: Joseph Reader & Charles Clark, '1932, a watershed year in nuclear physics', *Physics Today*, vol. 66, no. 3, March 2013, pp. 44–9, at p. 44.

43 'He felt ... that practical uses of his beloved nuclear physics': Mark Oliphant, 'The *Cambridge* Year: The Seventh Cecil Eddy Memorial Oration, 1964', *The Radiographer*, vol. II, no. 2, 1964, pp. 6–9, at p. 9.

43 'to run his own show': as cited in Cockburn & Ellyard, *Oliphant*, p. 66.

Chapter 3

46 'There is no greater cause of that fear'; 'it might be a good thing for this world'; 'Airforces ought all to be abolished'; 'The only defence is offence': Stanley Baldwin, House of Commons Debates, 10 November 1932, <hansard.millbanksystems.com/commons/1932/nov/10/international-affairs#S5CV0270P0_19321110_HOC_284>, accessed 15 November 2021.

48 'I fear war more than Fascism': as cited in Piers Brendon, *The Dark Valley: A Panorama of the 1930s*, Jonathan Cape, London, 2000, p. 354.

48 King George V threatened to join demonstrators: Brendon, *The Dark Valley*, p. 354.

50 'a certain openness among the public': SS Swords, *Technical History of the Beginnings of Radar*, Institution of Engineering and Technology, London, 2008, p. 75.

50 'chief of the militarists'; 'bring down airplanes': *New York Times*, 29 May 1924.

51 'Poor panic-stricken hordes': Siegfried Sassoon, 'Thoughts in 1932'; <www.poetrynook.com/poem/thoughts-1932>, accessed 26 April 2022.

52 'not peace' but 'an armistice for 20 years': <www.oxfordreference.com/view/10.1093/acref/9780191843730.001.0001/q-oro-ed5-00004492>, accessed 26 April 2022.

52 'unless science can find some way to come to the rescue': as cited in David E Fisher, *A Race on the Edge of Time: Radar – The Decisive Weapon of World War II*, McGraw-Hill, New York, 1989, p. 26; see also Rowe's account of the time, AP Rowe, *One Story of Radar*, Cambridge University Press, Cambridge, 1948.

52 'consider how far recent advances in scientific and technical knowledge': Terms of Reference for the Committee for the Scientific Survey of Air Defence, as cited in Swords, *Technical History*, p. 174.

Sources

53 'Suppose, just suppose ... that you had eight pints of water': Tim Harford, 'The search for a "death ray" led to radar', *50 Things that Made the Modern Economy*, 9 October 2017, radio broadcast, BBC world service, <www.bbc.com/news/business-41188464>, accessed 15 November 2021.

54 'attention is being turned': Memorandum from Robert Watson-Watt to the Committee for the Scientific Survey of Air Defence, January 1935, as cited in Swords, *Technical History*, p. 281.

54 'Detection and location of aircraft by radio methods': Memorandum from Robert Watson-Watt to the Committee for the Scientific Survey of Air Defence, 27 February 1935, as cited in Swords, *Technical History*, p. 281.

55 'electrical waves': Swords, *Technical History*, p. 2.

57 'Give them the third best'; 'cult of the imperfect': Robert Watson-Watt, *Three Steps to Victory: A Personal Account By Radar's Greatest Pioneer*, Odham's Press, London, 1957, p. 74.

58 'If you introduce that thing': as cited in Robert O'Connell, *Soul of the Sword: An Illustrated History of Weaponry and Warfare from Prehistory to the Present*, Free Press, New York, 2002, p. 292.

58 German research was transferred to the navy, where it 'languished': William Manchester & Paul Reid, *The Last Lion*, vol. 3, *Winston Spencer Churchill – Defender of the Realm, 1940–1965*, Little, Brown & Company, New York, 2012, p. 137; See Louis Brown, *Technical and Military Imperatives: A Radar History of World War II*, Institute of Physics Publishing, Bristol, 1999, pp. 33-96.

58 'the extent to which we had turned our discoveries': Winston Churchill, *The Second World War*, vol. 1, *The Gathering Storm*, Houghton Mifflin, Boston, 1948, p. 156.

59 'invisible bastion': Manchester & Reid, *The Last Lion*, vol. 3, p. 137.

61 Churchill knew about radar: William Manchester, *The Last Lion*, vol. 2, *Winston Spencer Churchill – Alone, 1932–1940*, Dell, New York, 1989, p. 569.

61 Churchill appointed to the Tizard Committee: Walter Kaiser, 'A case study in the relationship of history of technology and of general history: British radar technology and Neville Chamberlain's appeasement', *Icon*, vol. 2, 1996, pp. 29–52, at p. 36.

61 'a brandy – a big one'; 'evident relish': EG Bowen, *Radar Days*, Hilger, Bristol, 1987, p. 73.

65 The mould and its furry growth: Florey et al., *Antibiotics*, vol. 1, Oxford University Press, London, 1949, pp. 1–3; Milton Wainwright, 'Moulds in Folk Medicine', *Folklore*, vol. 100, no. 2, 1989, pp 162–6.

65	Indigenous Australians scraped mould: Lennard Bickel, *Rise up to Life: A Biography of Howard Walter Florey Who Made Penicillin and Gave It to the World*, Angus & Robertson, London, 1972, p. 61.
66	'play[ing] with microbes': Eric Lax, *The Mold in Dr. Florey's Coat: The Story of the Penicillin Miracle*, Henry Holt & Company, New York, 2004, p. 11.
66	'it is impossible': Lax, *The Mold in Dr. Florey's Coat*, p. 16.
67	'The trouble of making it seemed not worth while': as cited in Florey et al., *Antibiotics*, vol. 2, Oxford University Press, London, 1949, p. 633.
68	*Wings over Europe* play quotes: as cited in Charles A Carpenter, 'A "dramatic extravaganza" of the projected atomic age: *Wings over Europe*', *Modern Drama*, vol. 35, no. 4, 1992, pp. 552–61.
69	'Let it be split'; 'The Atom Split': as cited in Brian Cathcart, *The Fly in the Cathedral: How a Group of Cambridge Scientists won the International Race to Split the Atom*, Farrar, Straus & Giroux, New York, 2004, p. 249.
70	'play[ing] marbles': as cited in John Campbell, 'Rutherford, transmutation and the proton', *CERN Courier*, 8 May 2019, <cerncourier.com/a/rutherford-transmutation-and-the-proton/>, accessed 20 January 2022.
70	Rutherford called the prospect 'moonshine': Richard Rhodes, *The Making of the Atomic Bomb*, Simon & Schuster, New York, 1986, pp. 13–28.
71	'a wave of atomic disintegration': as cited in Diana Preston, *Before the Fall-Out: The Human Chain Reaction from Marie Curie to Hiroshima*, Corgi Books, London, 2006, p. 60.
71	'We may say that we are living': as cited in Robert Jungk, *Brighter than a Thousand Suns*, Harcourt, New York, 1958, p. 8.
71	'match to set the bonfire alight'; 'The discovery and control': Winston Churchill, 'Fifty Years Hence', *Macleans*, 15 November 1931, <archive.macleans.ca/article/1931/11/15/fifty-years-hence>, accessed 14 November 2021.
72	'Explosive forces, energy, materials': Churchill, 'Fifty Years Hence'.
75	'As the light changed to green': as cited in Rhodes, *The Making of the Atomic Bomb*, p. 28.
77	the energy released from splitting just one uranium atom: Ronald W Clark, *The Birth of the Bomb: The Untold Story of Britain's Part in the Weapon that Changed the World*, Phoenix House, London, 1961, p. 15.
79	'our present knowledge makes it seem possible': as cited in Preston, *Before the Fall-Out*, p. 202.
79	'Some physicists think': as cited in CP Snow, *The Physicists*, Macmillan, London, 1981, Appendix 1 'A new means of destruction' p. 176.
80	'Yet it must be made', Snow, *The Physicists*, p. 177.

Sources

80 'One was working against the clock': Robyn Williams, 'Sir Mark Oliphant', *Life and Times*, 1985, radio program, ABC Radio National, <www.abc.net.au/rn/legacy/programs/lifeandtimes/stories/2009/2588019.htm>, accessed 11 November 2021.

Chapter 4

82 one of only three overseas places listed by the Australian Government: Department of Agriculture, Water and Environment, *List of Overseas Places of Historic Significance to Australia*, <www.awe.gov.au/parks-heritage/heritage/places/list-overseas-places-historic-significance-australia>, accessed 15 November 2021.

83 'a mausoleum': Gwyn Macfarlane, *Howard Florey: The Making of a Great Scientist*, Oxford University Press, Oxford, 1979, p. 231.

83 'uneasily aware of their own slackness': Macfarlane, *Howard Florey*, p. 239.

84 'much relieved': as cited in Eric Lax, *The Mold in Dr. Florey's Coat: The Story of the Penicillin Miracle*, Henry Holt & Company, New York, 2004, p. 57.

84 'Perhaps the most useful lesson': H Florey, 'Penicillin', *Nobel Lecture*, 11 December 1945, <www.nobelprize.org/prizes/medicine/1945/florey/lecture/>, accessed 15 November 2021.

84 Chain fled to Britain when the Nazis came to power: Ronald W Clark, *The Life of Ernst Chain: Penicillin and Beyond*, Weidenfeld & Nicolson, London, 1985, p. 8.

86 'the very walls of Florey's office would shudder': as cited in Lennard Bickel, *Rise up to Life: A Biography of Howard Walter Florey Who Made Penicillin and Gave It to the World*, Angus & Robertson, London, 1972, p. 49.

87 'A good sample will completely inhibit staphylococci': Alexander Fleming, 'On the antibacterial action of cultures of a penicillium, with special reference to their use in the isolation of *B. Influenzae*', *British Journal of Experimental Pathology*, vol. 10, no. 3, June 1929, pp. 226–36, at pp. 232, 235.

87 Chain's interest was immediate: Macfarlane, *Howard Florey*, p. 281; Clark, *The Life of Ernst Chain*, p. 37.

87 'something seemed to click': Clark, *The Life of Ernst Chain*, p. 33.

88 'This is the tree': Max Blythe, 'Dr Paquita McMichael in interview with Dr Max Blythe, Edinburgh, 17 February 1998', Royal College of Physicians and Oxford Brookes University Medical Sciences Video Archive MSVA 178, <radar.brookes.ac.uk/radar/file/6b6a8fce-fa49-45f9-935a-6394f699f654/1/McMichael%2CP.pdf>, accessed 21 November 2021.

88 'There is no question we will now have to go for penicillin': as cited in Bickel, *Rise up to Life*, p. 66.

89 What swayed Florey was his familiarity with the substance: Macfarlane, *Howard Florey*, p. 282; see also Ronald P Rubin, 'A brief history of great discoveries in pharmacology: In celebration of the centennial anniversary of the founding of the American Society of Pharmacology and Experimental Therapeutics', *Pharmacological Review*, vol. 59, no. 4, 2007, pp. 289–359, at p. 321.

90 Florey's role as fundraiser: Lax, *The Mold in Dr Florey's Coat*, pp. 82–4.

90 'fed up' with the 'difficulties of trying to keep work going here': as cited in Bickel, *Rise up to Life*, p. 71.

90 Florey and Chain were motivated more by scientific interest: Macfarlane, *Howard Florey*, p. 285; for a different view see Clark, *The Life of Ernst Chain*, p. 41.

91 the first manufacturing town in the world: Eric Hopkins, *Birmingham: The First Manufacturing Town in the World, 1760–1840*, Weidenfeld & Nicolson, London, 1989.

92 the first successful case of cancer radiation therapy: Bill Bryson, *The Body: A Guide for Occupants*, Penguin, London, 2020, p. 402.

92 Oliphant would later consult Florey: Clarence Larson, *Pioneers of Science and Technology: Interview with Mark Oliphant*, Washington DC, 11 March 1988, <www.youtube.com/watch?v=z0QaXvLsR9A>, accessed 10 November 2021.

93 'a major player in the drama of the irruption of nuclear physics': Sam Edwards, 'Rudolph E Peierls', *Physics Today*, vol. 49, no. 2, 1996, p. 74.

93 'What would you say': M Shifman, 'The Beginning of the Nuclear Age', pp. 1–35, at p. 12, <www.semanticscholar.org/paper/The-Beginning-of-the-Nuclear-Age-Shifman/e10709fc56d91ac84b5ff50b1ec5ed320d46de07>, accessed 25 January 2022.

94 Oliphant contemplated the terrible possibility of an atomic bomb: Stewart Cockburn & David Ellyard, *Oliphant: The Life and Times of Sir Mark Oliphant*, Axiom Books, Adelaide, 1981, p. 96.

95 'You just come over in the summer': Charles Weiner, 'Oral histories: Otto Frisch', Niels Bohr Library and Archives, American Institute of Physics, 3 May 1967, <www.aip.org/history-programs/niels-bohr-library/oral-histories/4616>, accessed 15 November 2021.

95 'I had a fear ... that Denmark would soon be overrun by Hitler': Weiner, 'Oral histories: Otto Frisch'.

95 'There was very little else to do': Otto Frisch, *What Little I Remember*, Cambridge University Press, New York, 1979, p. 121.

96 'all was overtaken by the war': Mark Oliphant, 'Looking back: Sir Mark Oliphant in conversation with David Ellyard', in Macfarlane Burnet & Mark

Sources

Oliphant, *Sir Frank Macfarlane Burnet on Ageing & Looking Back by Sir Mark Oliphant*, Australian Broadcasting Commission, Sydney, 1979, p. 37.

96 'a man of action rather than of ideas': Mark Oliphant & Lord Penney, 'John Douglas Cockcroft, 1897–1967', *Biographical Memoirs of Fellows of the Royal Society*, vol. 14, November 1968, p. 143.

97 'These devices would be troublesome': as cited in Stephen Phelps, *The Tizard Mission: The Top-Secret Operation that Changed the Course of World War II*, Westholme, Yardley, 2010, p. 45.

97 Oliphant readily understood the problem: Robert Buderi, *The Invention that Changed the World: The Story of Radar from War to Peace*, Little, Brown & Company, London, 1997, p. 83.

98 'I did a back-of-an-envelope calculation': EG Bowen, *Radar Days*, Hilger, Bristol, 1987, p. 143.

99 'there was practically no interest within Bawdsey Manor': Bowen, *Radar Days*, p. 143.

100 'think in fundamental terms rather than just follow radio practice': Robin Hughes, *Australian Lives: Stories of Twentieth Century Australians*, Angus & Robertson, Sydney, 1996, p. 127.

101 'By now ... we were all interested'; was 'most exciting, but apparently impossible': HAH Boot & JT Randall, 'Historical notes on the cavity magnetron', *IEEE Transactions on Electron Devices*, vol. 23, no. 7, 1976, pp. 724–9, at p. 724.

Chapter 5

102 'No conqueror returning': as cited in AN Wilson, *After the Victorians: 1901–1953*, Hutchinson, London, 2005, p. 365.

103 '*4.15 Uhr geht es los. Gott sei dank*' [At 4:15 am we begin]: '1 Września 1939 r., Wieluń. Masakra na bezbronnym mieście', *Interia*, 1 September 2019, <wydarzenia.interia.pl/raporty/raport-zbrodnia-bez-kary/historie/news-1-wrzesnia-1939-r-wielun-masakra-na-bezbronnym-miescie,nId,3179824>, accessed 24 November 2021.

104 'I should like the gentlemen of London': as cited in 'Luftwaffe Air War Poland 1939', *Weapons and Warfare*, 4 May 2019, <weaponsandwarfare.com/2019/05/04/luftwaffe-air-war-poland-1939/>, accessed 24 November 2021.

105 'shocked silence': as cited in Wilson, *After the Victorians*, p. 378.

105 Military planners had estimated 20000 casualties: David Reynolds, *The Long Shadow: The Great War and the Twentieth Century*, Simon & Schuster, London, 2013, p. 225.

105 'The British public must not be led to expect': WK Hancock, *Country and Calling*, Faber & Faber, London, 1954, p. 184.
105 'When the sun is low': as cited in Norman Longmate, *How We Lived Then: A History of Everyday Life During the Second World War*, London, 2002, p. 32.
106 television engineers redeployed to work on radar: 'Close down of television service for the duration of the war, 1 September 1939', History of the BBC, BBC website, <www.bbc.com/historyofthebbc/anniversaries/september/closedown-of-television>, accessed 24 November 2021.
108 'All the sensations of war': AJP Taylor, *The Second World War: An Illustrated History* [1975] in *The Second World War and Its Aftermath*, Folio Society, London, 1998, p. 36.
108 'And not only outsiders, but colonials!': David E Fisher, *A Race on the Edge of Time: Radar – The Decisive Weapon of World War II*, McGraw Hill, New York, 1989, p. 249.
108 'give us many watts': Watson-Watt, *Three Steps to Victory: A Personal Account by Radar's Greatest Pioneer*, Odhams Press, London, 1957, p. 287.
109 Early magnetrons: James E Brittain, 'The magnetron and the beginnings of the microwave age', *Physics Today*, vol. 38, no. 7, July 1985, pp. 60–7; Yves Blanchard, Gaspare Galati & Pietvan Genderen, 'The cavity magnetron: Not just a British invention', *IEEE Antennas and Propagation Magazine*, vol. 55, no. 5, October 2013, pp. 244–54; Allison Marsh, 'From World War II radar to microwave popcorn, the cavity magnetron was there', *IEEE Spectrum*, 31 October 2018, <spectrum.ieee.org/magnetron>, accessed 2 December 2021.
110 'Fortunately we did not have the time to survey': HAH Boot & JT Randall, 'Historical notes on the cavity magnetron', *IEEE Transactions on Electron Devices*, vol. 23, no. 7, 1976, pp. 724–29, at p. 724.
110 the klystron and the magnetron might be combined: Ronald W Clark, *The Rise of the Boffins*, Phoenix House, London, 1962, p. 130.
112 'The Magnetron Memorandum'; 'a communal manifesto': Watson-Watt, *Three Steps to Victory*, p. 289.
115 'great drive'; 'fortunate ... that we were in Oliphant's laboratory': Boot & Randall, 'Historical notes on the cavity magnetron', p. 724.
116 'At one period the ordinary empty soft drink bottle': Norman Heatley, 'Penicillin and luck', in Carol L Moberg & Zanvil A Cohn (eds), *Launching the Antibiotic Era: Personal Accounts of the Discovery and Use of the First Antibiotics*, Rockefeller University Press, New York, 1990, pp. 31–41, at p. 33.
116 'Without Fleming, no Chain': Henry Harris, 'Howard Florey and the development of penicillin', *Notes and Records: The Royal Society Journal of the History of Science*, vol. 53, no. 2, 1999, pp. 243–52, at p. 249.

Sources

117 Photograph of Heatley digging air-raid shelter: Eric Lax, *The Mold in Dr. Florey's Coat: The Story of the Penicillin Miracle*, Henry Holt & Company, New York, 2004, photos between pp. 148 & 149.

117 Florey, a keen photographer, recorded much of it: Edward Abraham, 'Oxford, Howard Florey and World War II' in Moberg & Cohn (eds), *Launching the Antibiotic Era*, pp. 19–30, at p. 20; see also: Robert Root-Bernstein, 'Howard Florey: Photographer, cinematographer and Sunday painter', *Leonardo*, vol. 42, no. 3, 2009, p. 265.

118 'Chain's enthusiasm overcame Florey's scientific caution': Gwyn Macfarlane, *Howard Florey: The Making of a Great Scientist*, Oxford University Press, Oxford, 1979, p. 300.

118 'Penicillin can easily be prepared in large amounts'; 'antiseptic action' directly '*in vivo*': Macfarlane, *Howard Florey*, p. 299.

119 Florey reached out for help to Rockefeller Foundation: Gerald Jonas, *The Circuit Riders: Rockefeller Money and the Rise of Modern Science*, WW Norton, New York, 1989, pp. 230ff.

120 Harry M Miller's diaries: Memorandum 'HMM to WW, Paris No 36: Visits with British scientists in Oxford, November 1, 1939', 6 November 1939, Rockefeller Archive Centre.

120 'just on the point'; 'practically the only experimental pathologist'; 'maximum would certainly not exceed 1500 pounds'; 'should be given serious study': Jonas, *The Circuit Riders*, pp. 267–8.

121 'It may also be pointed out that the work proposed': Jonas, *The Circuit Riders*, p. 269.

121 'royal generosity': Ernst Chain, 'Thirty years of penicillin therapy', *Proceedings of the Royal Society of London*, vol. 179 (Series B), 1971, pp 293–319, at p. 298.

124 'the crucial day in the whole development of penicillin': Ronald W Clark, *The Life of Ernst Chain: Penicillin and Beyond*, Weidenfeld & Nicolson, London, 1985, p. 48.

125 'absolutely a visionary': Charles Weiner, 'Oral histories: Otto Frisch', Niels Bohr Library and Archives, American Institute of Physics, 3 May 1967, <www.aip.org/history-programs/niels-bohr-library/oral-histories/4616>, accessed 15 November 2021.

125 'I am told that you have refugees'; 'In my opinion it is much more important': Ronald W Clark, *Tizard*, Methuen, London, 1965, p. 214.

126 'Peierls knew that this was connected with the generation of very short electric waves': Otto Frisch, *What Little I Remember*, Cambridge University Press, New York, 1979, p. 123.

126 "so preposterous then to think of separating U-235': John Wheeler, 'Mechanism of Fission', *Physics Today*, November 1967, vol. 20, no. 11, pp. 49–52, at p. 52.
127 'of the order of tons': Richard Rhodes, *The Making of the Atomic Bomb*, Simon & Schuster, New York, 1986, p. 321.
127 'saw no reason against having my paper published': Rudolf Peierls, *Bird of Passage: Recollections of a Physicist*, Princeton University Press, Princeton, 1985, p. 153.
128 'there may be no possibility missed'; 'I understand … that you do not anticipate': as cited in Andrew Ramsey, *The Basis of Everything: Rutherford, Oliphant and the Coming of the Atomic Bomb*, Harper Collins, Sydney, 2019, p. 220.
128 'the construction of a super bomb': as cited in Margaret Gowing, *Britain and Atomic Energy, 1939–1945*, Macmillan, London, 1964, p. 40.
128 'no worse than setting fire to a similar quantity': Frisch, *What Little I Remember*, p. 125.
129 'something like a pound or two': Frisch, *What Little I Remember*, p. 126.
129 'Suppose someone gave you a quantity of pure 235 isotope'; 'order of magnitude was right': Peierls, *Bird of Passage*, p. 154.
130 'the equivalent of thousands of tons'; 'We were quite staggered'; 'an atomic bomb was possible': Peierls, *Bird of Passage*, p. 154.
130 'the jitterbug': Sue Rabbitt Roff, 'Making the jitterbug work: Marcus Oliphant and the Manhattan Project', 30 May 2019, <www.atomicheritage.org/history/making-jitterbug-work-marcus-oliphant-and-manhattan-project>, accessed 6 January 2022.
130 'Any competent nuclear physicist': as cited in Ronald W Clark, *The Birth of the Bomb: The Untold Story of Britain's Part in the Weapon that Changed the World*, Phoenix House, London, 1961, p. 51.
130 'We were at war, and the idea was reasonably obvious': Frisch, *What Little I Remember*, p. 126.
130 'Look … shouldn't somebody know about that?': Charles Weiner, 'Oral histories: Otto Frisch'.
130 'how to send a secret communication': Peierls, *Bird of Passage*, p. 155.
131 'On the Construction of a "Super-bomb": based on a Nuclear Chain Reaction in Uranium': reproduced in Gowing, *Britain and Atomic Energy*, Appendix 1, pp. 389–93.
132 'Memorandum on the Properties of a Radioactive "Super-bomb"': reproduced in Clark, *Tizard*, pp. 214–17.
133 'In say 1937 everybody was horrified': 'On the right side of wrong', 17 February 2000, television documentary, *BBC (Midlands)*.

Sources

133 'had caught an elephant in the jungle by the tail': as cited in Sabina Lee, 'Birmingham – London – Los Alamos – Hiroshima: Britain and the Atomic Bomb', *Midland History*, vol. 21, no. 1, 2002, pp. 146–64, at p. 152.
134 'street-wise': Lorna Arnold, 'The history of nuclear weapons: The Frisch–Peierls memorandum on the possible construction of atomic bombs of February 1940', *Cold War History*, vol. 3, no. 3, 2003, pp. 111–26, at p. 113.
135 Tizard was a sceptic: Gowing, *Britain and Atomic Energy*, p. 35.
135 'I have considered these suggestions in some detail': as cited in Clark, *Tizard*, p. 218.
136 'What I should like': as cited in Clark, *Tizard*, p. 218.

Chapter 6

137 'Mr Churchill's sun': as cited in Fredrik Logevall, *JFK*, vol. 1, *1917–1956*, Viking, London, 2020, p. 257.
139 'I only wish the position had come your way in better times': as cited in John Lukacs, *Five Days in London: May 1940*, Scribe, Sydney, 2001, p. 6.
139 Hitler, who danced a jig of joy: William Manchester & Paul Reid, *The Last Lion*, vol. 3: *Winston Spencer Churchill – Defender of the Realm, 1940–1965*, Little, Brown & Company, New York, 2012, p. 49.
140 Fuelled by Romanian and Soviet oil, steely determination and methamphetamines: Norman Ohler, *Blitzed: Drugs in Nazi Germany*, Penguin, London, 2017, pp. 77–9, 84ff.
140 'At the moment it looks like the greatest military disaster': as cited in Lukacs, *Five Days in London*, p. 18.
140 'We shall have lost all our trained soldiery': as cited in Manchester & Reid, *The Last Lion*, vol. 3, p. 81.
143 Unknown even to Allier: Per F Dahl, *Heavy Water and the Wartime Race for Nuclear Energy*, Institute of Physics Publishing, Bristol, 1999, pp. 103–8.
143 MAUD Committee name: Ronald W Clark, *The Birth of the Bomb: The Untold Story of Britain's Part in the Weapon that Changed the World*, Phoenix House, London, 1961, pp. 76–7; Margaret Gowing, *Britain and Atomic Energy, 1939–1945*, Macmillan, London, 1964, p. 45; Graham Farmelo, *Churchill's Bomb: How the United States Overtook Britain in the First Nuclear Arms Race*, Basic Books, New York, 2013, p. 161.
146 At precisely 11 am: the experiment is recounted in HW Florey et al., *Antibiotics*, vol. 2, Oxford University Press, London, 1949, p. 638.
147 Florey recorded that the treated mice were not showing any distress: Lennard Bickel, *Rise up to Life: A Biography of Howard Walter Florey Who Made Penicillin and Gave It to the World,*, Angus & Robertson, London, 1972, p. 96.

147 'nearly dead. The others in a poor way': as cited in Eric Lax, *The Mold in Dr. Florey's Coat: The Story of the Penicillin Miracle*, Henry Holt & Company, New York, 2004, p. 119.
148 'At three minutes to midnight': as cited in Bickel, *Rise up to Life*, p. 97.
148 'relief, joy, happiness': as cited in Lax, *The Mold in Dr Florey's Coat*, p. 120.
149 'the blackest day of all': as cited in Roy Jenkins, *Churchill*, Macmillan, London, 2001, p. 603.
149 'What a grim interlude': Harold Nicolson, *Diaries and Letters: 1939–1945*, Collins, London, 1967, p. 25.
149 'ate and drank with evident distaste': as cited in Jenkins, *Churchill*, p. 604.
150 'It seems like a miracle': as cited in Bickel, *Rise up to Life*, p. 98.
150 'Adolf Hitler came closest': John Lukacs, *Five Days in London*, p. 1.
150 'Chain was beside himself with excitement': as cited in Brenda Heagney, *Half a Century of Penicillin: An Australian Perspective*, Royal Australasian College of Physicians, Sydney, 1991, p. 4.
150 'pure' brown powder was more than 99.5 per cent 'junk': Lax, *The Mold in Dr Florey's Coat*, p. 119; Gwyn Macfarlane, *Howard Florey: The Making of a Great Scientist*, Oxford University Press, Oxford, 1979, p. 317.
151 'border[ed] on the miraculous': Ernst Chain, "Thirty years of penicillin therapy"', *Proceedings of the Royal Society of London*, vol. 179 (Series B), 1971, pp. 293–319, at p. 304.
152 weighs 3000 times as much as a mouse: Macfarlane, *Howard Florey*, p. 316.
153 Churchill thought 30 000 troops might be rescued: Joshua Levine, *Dunkirk: The History Behind the Major Motion Picture*, William Collins, London, 2017, p. 216.
153 'There was a white glow': Winston Churchill, *The Second World War*, vol. 2, *Their Finest Hour*, Houghton Mifflin, Boston, 1949, p. 100.
153 'You and I will be dead in three months' time': as cited in Norman Rose, *Churchill: An Unruly Life*, Tauris Parke, New York, 2009, p. 264.
154 'Wars are not won by evacuations': Winston Churchill, House of Commons Debates, 4 June 1940, <api.parliament.uk/historic-hansard/commons/1940/jun/04/war-situation>, accessed 17 November 2021.
156 'We have been working with a substance': as cited in Lax, *The Mold in Dr Florey's Coat*, p. 122.
157 '"The Battle of France" is over': Winston Churchill, House of Commons Debates, 18 June 1940, <api.parliament.uk/historic-hansard/commons/1940/jun/18/war-situation>, accessed 17 November 2021.
157 Nazi soldiers sang 'We're sailing against England': Manchester & Reid, *The Last Lion*, vol. 3, p. 113.

Sources

158 'I shall not go down like the others': as cited in Manchester & Reid, *The Last Lion*, vol. 3, p. 124.
158 'their finest hour': Winston Churchill, House of Commons Debates, 18 June 1940, <api.parliament.uk/historic-hansard/commons/1940/jun/18/war-situation>, accessed 17 November 2021.
158 'terrible beauty': Jenkins, *Churchill*, p. 611.
158 'They expect an invasion this weekend': Nicolson, *Diaries and Letters*, p. 101.
160 'Penicillin as a chemotherapeutic agent': E Chain, HW Florey, AD Gardner, NG Heatley, MA Jennings, J Orr-Ewing & AG Sanders, *Lancet*, vol. 236, no. 6104, 24 August 1940, pp. 226–8.
160 'In recent years interest in chemotherapeutic effects': Chain et al., 'Penicillin as a chemotherapeutic agent', p. 226.
161 'During the last year methods have been devised here': Chain et al., 'Penicillin as a chemotherapeutic agent', p. 227.
161 'The results are clear cut': Chain et al., 'Penicillin as a chemotherapeutic agent', p. 228.
161 'What its chemical nature is': as cited in Macfarlane, *Howard Florey*, p. 322.
163 'Good God! ... I thought he was dead': as cited in Bickel, *Rise up to Life*, p. 110.
163 'I hear you've been doing things': as cited in Bickel, *Rise up to Life*, p. 110.
164 'What reserves have we?': as cited in Rose, *Churchill*, p. 265.
164 115 new Hurricanes and Spitfires: Manchester & Reid, *The Last Lion*, vol. 3, p. 149.
164 'Never in the field of human conflict': Winston Churchill, House of Commons Debates, 20 August 1940, <api.parliament.uk/historic-hansard/commons/1940/aug/20/war-situation >, accessed 17 November 2021.
165 'one of the greatest combined feats': HV Jones, 'Sir Robert Alexander Watson-Watt', *Oxford Dictionary of National Bibliography*, <www.oxforddnb.com/view/10.1093/ref:odnb/9780198614128.001.0001/odnb-9780198614128-e-9000022?rskey=NMgm8z&result=1>, accessed 12 January 2022.
166 eat more carrots; recruiting cats: Manchester & Reid, *The Last Lion*, vol. 3, p. 269.
166 'critical mistake': John T Correll, 'How the Luftwaffe lost the battle of Britain', *Air Force Magazine*, 1 August 2008, <www.airforcemag.com/article/0808battle/>, accessed 18 November 2021.
168 'More than once ... I had to report lights being shown from Oliphant's lab'; Stewart Cockburn, 'Keith Hancock interviewed by Stewart Cockburn for the Mark Oliphant biography collection', 14 February 1980, National Library of Australia, ORAL TRC 889.

Chapter 7

170 Taffy Bowen's journey with the black box: EG Bowen, *Radar Days*, Hilger, Bristol, 1987, pp. 153ff.
171 'precious magnetron': as cited in Stephen Phelps, *The Tizard Mission: The Top-Secret Operation that Changed the Course of World War II*, Westholme, Yardley, 2010, p. 143.
171 'briefcase that changed the world': Angela Hind, 'Briefcase that "changed the world"', *The World in a Briefcase*, 5 February 2007, radio program, Pier Productions, BBC Radio 4, <news.bbc.co.uk/2/hi/science/nature/6331897.stm>, accessed 18 November 2021.
173 'if things go really badly with this country': as cited in Diana Preston, *Before the Fall-Out: The Human Chain Reaction from Marie Curie to Hiroshima*, Corgi Books, London, 2006, p. 237.
176 'a sign ... that we didn't know what to do next': as cited in John S Rigden, *Rabi: Scientist and Citizen*, Basic Books, New York, 1987, p. 129.
176 'conspicuously drunk': Jennet Conant, *Tuxedo Park: A Wall Street Tycoon and the Secret Palace of Science that Changed the Course of World War II*, Simon & Schuster, New York, 2003, p. 190.
176 the Americans were 'shaken': Bowen, *Radar Days*, p. 159.
176 The 'crucial' meeting: Bowen, *Radar Days*, p. 159.
177 'devious charm': Conant, *Tuxedo Park*, p. 200.
177 'increased the power available': as cited in Conant, *Tuxedo Park*, p. 191.
177 'The atmosphere was electric': Bowen, *Radar Days*, p. 162.
178 'Alfred Loomis came in in the afternoon': as cited in Conant, *Tuxedo Park*, p. 193.
179 Lawrence the 'godfather'; 'No one else in American physics': Watson-Watt, *Three Steps to Victory: A Personal Account by Radar's Greatest Pioneer*, Odhams Press, London, 1957, p. 297.
179 'If Lawrence was interested in the program': as cited in Conant, *Tuxedo Park*, p. 202.
179 'it was a remarkable invention': Luis W Alvarez, *Adventures of a Physicist*, Basic Books, New York, 1987, p. 83.
179 'model to find resonant frequencies of the O. Tube': as cited in Raphael Chayim Rosen, '*Under the radar: Physics, engineering, and the distortion of a World War Two legacy*', honors thesis, Harvard University, Cambridge, Mass., March 2006, p. 29, note 84, <canvas.harvard.edu/files/4329530/download?download_frd=1>, accessed 18 November 2021.

Sources

180 'what you are after is to see that the Nazis don't blow us up': as cited in Richard Rhodes, *The Making of the Atomic Bomb*, Simon & Schuster, New York, 1986, p. 314.

181 The Uranium Committee inaction: Phelps, *The Tizard Mission*, pp. 282–3; Rhodes, *The Making of the Atomic Bomb*, pp. 315–7, 336–8.

182 'no possibility within practical range'; 'sheer waste of time': quoted in David Zimmerman, *Top Secret Exchange: The Tizard Mission and the Scientific War*, Alan Sutton Publishing Limited, Phoenix Mill, 1996, p. 151.

182 a 'man whose experience combined': Ronald W Clark, *The Birth of the Bomb: The Untold Story of Britain's Part in the Weapon that Changed the World*, Phoenix House, London, 1961, p. 161.

182 The Tizard Mission and the atom bomb: David Zimmerman, 'The Tizard Mission and the development of the atomic bomb', *War in History*, vol. 2, no. 3, 1995, pp. 259–73, at p. 267.

183 'Be careful': as cited in Leslie R Groves, *Now It Can Be Told: The Story of the Manhattan Project*, Andre Deutsch, London, 1963, p. 33.

183 uranium in a Staten Island warehouse: Groves, *Now It Can Be Told*, p. 34.

184 'that the uranium investigation in North America': Margaret Gowing, *Britain and Atomic Energy, 1939–1945*, Macmillan, London, 1964, p. 65.

184 'have damned little to offer': quoted in Graham Farmelo, *Churchill's Bomb: How the United States Overtook Britain in the First Nuclear Arms Race*, Basic Books, New York, 2013, p. 167.

184 representatives of Bush's NDRC: Zimmerman, 'The Tizard Mission', pp. 271–2.

184 the official US history: Richard G Hewlett & Oscar E Anderson, Jr. *History of the United States Atomic Energy Commission*, vol. 1, *The New World, 1939–1946*, Pennsylvania State University Press, University Park, 1962, p. 40.

185 not a single member of the Briggs Committee: Arthur H Compton, *Atomic Quest: A Personal Narrative*, Oxford University Press, London, 1956, p. 46.

186 'had thought a great deal about the problems': Rudolf Peierls, *Bird of Passage: Recollections of a Physicist*, Princeton University Press, Princeton, 1985, p. 155.

187 'By March 1941 therefore an atomic bomb': Gowing, *Britain and Atomic Energy*, p. 68.

187 'was not only possible – it was inevitable': as cited in Farmelo, *Churchill's Bomb*, p. 179.

187 'was feasible': as cited in Clark, *The Birth of the Bomb*, p. 137.

188 'in England or at worst in Canada': as cited in Farmelo, *Churchill's Bomb*, p. 188.

188 'Although personally I am quite content': as cited in Farmelo, *Churchill's Bomb*, p. 190.

188 'some lonely, uninhabited island': as cited in Farmelo, *Churchill's Bomb*, p. 190.
188 British versus US spending: Gowing, *Britain and Atomic Energy*, p. 162.
190 'The yield was pathetically low': Norman Heatley, 'Penicillin and luck', in Carol L Moberg & Zanvil A Cohn (eds), *Launching the Antibiotic Era: Personal Accounts of the Discovery and Use of the First Antibiotics*, Rockefeller University Press, New York, 1990, pp. 36–7.
191 'out came the pocket knife': Ruth Richardson, 'Heatley's Vessel', *Lancet*, vol. 357, no. 9264, 21 April 2001, p. 1298.
191 'nearly one thousand times': as cited in Eric Lax, *The Mold in Dr. Florey's Coat: The Story of the Penicillin Miracle*, Henry Holt & Company, New York, 2004, p. 149.
192 Dr Martin Dawson began injecting patients: Gladys L Hobby, *Penicillin: Meeting the Challenge*, Yale University Press, New Haven, 1985, pp. 69–80.
193 Alexander was dying from a simple scratch to his face: Mike Barrett, 'Penicillin's first patient', *Mosaic*, 13 May 2018, <mosaicscience.com/story/penicillin-first-patient-history-albert-alexander-AMR-DRI/>, accessed 24 November 2021.
194 Alexander and the first clinical human tests: HW Florey et al., *Antibiotics*, Oxford University Press, London, 1949, vol. 2, pp. 646–8.
195 'almost miraculous': as cited in Gerald Jonas, *The Circuit Riders: Rockefeller Money and the Rise of Modern Science*, WW Norton, New York, 1989, p. 297.
195 'our greatest piece of luck': Heatley, 'Penicillin and luck', p. 41.
196 'I left the room silently': as cited in Ronald W Clark, *The Life of Ernst Chain: Penicillin and Beyond*, Weidenfeld & Nicolson, London, 1985, p. 68.

Chapter 8

198 'freely-lit, non-rationed paradise': as cited in Eric Lax, *The Mold in Dr. Florey's Coat: The Story of the Penicillin Miracle*, Henry Holt & Company, New York, 2004, p. 171.
198 with a spoonful or five of sugar in their coffee: 'The mould, the myth and the microbe', *Horizon*, 27 January 1986, television documentary, BBC.
199 'bristling with wire and guards': as cited in Lax, *The Mold in Dr. Florey's Coat*, p. 171.
199 'a Mr Makinsky': Lennard Bickel, *Rise up to Life: A Biography of Howard Walter Florey Who Made Penicillin and Gave It to the World*, Angus & Robertson, London, 1972, p. 138.
199 'suave, dapper': EJ Kahn, 'The universal drink: The making-and selling-of Coca-Cola', *New Yorker*, 14 February 1959, <www.newyorker.com/magazine/1959/02/14/the-universal-drink>, accessed 30 November 2021.

Sources

199 the most physically beautiful of Europe's dictators: Neill Lochery, *Lisbon: War in the Shadows of the City of Light, 1939–1945*, Scribe, Melbourne, 2011, p. 14.

200 Florey and his two cameras: Robert Root-Bernstein, 'Howard Florey: Photographer, cinematographer and Sunday painter', *Leonardo*, vol. 42, no. 3, 2009, p. 265. Nineteen of Florey's 16 mm films, including his trip to the USA in 1941, were deemed worthy of preservation by the National Film Archive of the British Film Institute and are stored in their archives in London.

200 a city that traded in two currencies: Jeremy Adelman, *Worldly Philosopher: The Odyssey of Albert O. Hirschman*, Princeton University Press, Princeton, 2013, p. 184.

200 'idle and solitary men': Bickel, *Rise up to Life*, p. 138.

201 the model for James Bond: Larry Loftis, *Into the Lion's Mouth: The True Story of Dusko Popov: World War II Spy, Patriot and the Real-life Inspiration for James Bond*, Berkley Caliber, New York, 2016.

202 'medical business; 'her great-uncle, Albert Moissan': *New York Times*, 3 July 1941.

203 'I remember him best of all for that performance': as cited in Bickel, *Rise up to Life*, p. 141.

204 'man with a carpet bag': 'Penicillin – Like lottery luck', *Courier Mail*, undated, Royal Society, M1944, Series 371, AJCP ref: 74_nla.obj-739916847-m.

205 NRRL found ways to increase penicillin yields: AN Richards, 'Production of penicillin in the United States (1941–1946)', *Nature*, no. 4918, 1 February 1964, pp. 441–5.

207 'I have literally met dozens of people': as cited in Lax, *The Mold in Dr. Florey's Coat*, p. 181.

207 'The mold is as temperamental as an opera singer': as cited in ME Bowden, 'Old brew, new brew', *Distillations*, Science History Institute, 30 July 2018, <www.sciencehistory.org/distillations/old-brew-new-brew>, accessed 19 November 2021.

207 in case penicillin could be synthesised: Marc Landas, *Cold War Resistance: The International Struggle over Antibiotics*, Potomac Books, Lincoln, 2020, p. 38.

208 hesitancy was proving 'the norm': Landas, *Cold War Resistance*, p. 38.

208 'It was the impression of the Committee': as cited in Landas, *Cold War Resistance*, p. 38.

209 'rough colonial genius': as cited in Bickel, *Rise up to Life*, p. 27.

209 Richards would back Florey: Richards, 'Production of penicillin in the United States (1941–1946)', p. 442.

209 Richards 'encouraged'; 'the difficulties which attended its production': Irvin Stewart, *Organizing Scientific Research for War: The Administrative History of the Office of Scientific Research and Development*, Atlantic-Little Brown, Boston, 1948, p. 105.

209 'Florey is a scientist': as cited in Bickel, *Rise up to Life*, p. 152.

210 'Without Richards, Americans would never have taken over production': as cited in Lax, *The Mold in Dr. Florey's Coat*, p. 187.

211 Churchill ordered his oxygen mask to be customised: William Manchester & Paul Reid, *The Last Lion*, vol. 3, *Winston Spencer Churchill – Defender of the Realm, 1940–1965*, Little, Brown & Company, New York, 2012, p. 556.

211 'Britain's hope lies in Russia and America': as cited in Manchester & Reid, *The Last Lion*, vol. 3, p. 119.

212 Americans expected the Soviet Union to resist: AJP Taylor, *The Second World War: An Illustrated History* [1975] in *The Second World War and Its Aftermath*, Folio Society, London, 1998, p. 86.

212 killing several hundred crew: 'Casualty Lists of the Royal Navy and Dominion Navies, World War 2. 1st–30th April 1941', <www.naval-history.net/xDKCas1941-04APR.htm>, accessed 28 April 2022.

Chapter 9

215 the importance of proximity fuzes: Stewart, *Organizing Scientific Research for War*, pp. 123–4; Vannevar Bush, 'Foreword', in James Phinney Baxter, *Scientists Against Time*, 2nd edn, MIT Press, Cambridge, Mass., 1968.

215 'Bugger it all, this is not really interesting work': as cited in Robin Hughes, *Australian Lives: Stories of Twentieth Century Australians*, Angus & Robertson, Sydney, 1996, p. 127.

216 'discreet inquiries': as cited in Graham Farmelo, *Churchill's Bomb: How the United States Overtook Britain in the First Nuclear Arms Race*, Basic Books, New York, 2013, p. 197.

216 'crossed the United States like a nuclear Paul Revere': McGeorge Bundy, *Danger and Survival: Choices about the Bomb in the First Fifty Years*, Vintage Books, New York, 1990, p. 49.

218 'amazed and distressed'; 'This inarticulate and unimpressive man': Mark Oliphant, 'The beginning: Chadwick and the neutron', *Bulletin of the Atomic Scientists*, vol. 38, no. 10, 1982, pp. 14–18, at p. 17.

218 'Here it was September 1941': Lennard Bickel, *The Deadly Element: The Story of Uranium*, Stein & Day, New York, 1979, p. 164.

Sources

218 'study blowing people up': Nuel Pharr Davis, *Lawrence and Oppenheimer*, Fawcett, New York, 1968, p. 110.

218 'Oliphantic': Farmelo, *Churchill's Bomb*, p. 196.

219 'said "bomb" in no uncertain terms' as cited in Richard Rhodes, *The Making of the Atomic Bomb*, Simon & Schuster, New York, 1986, p. 373.

219 'since being created in 1939': Davis, *Lawrence and Oppenheimer*, p. 110.

219 Leo Szilard argued later about delay: Davis, *Lawrence and Oppenheimer*, p. 112.

220 Conant perhaps the most influential university administrator since Wilson: Harold Taylor, 'What kind of person is James Bryant Conant? What has been his influence?', *New York Times*, 22 March 1970, p. 288, <www.nytimes.com/1970/03/22/archives/what-kind-of-person-is-james-bryant-conant-what-has-been-his.html>, accessed 12 January 2022.

220 'impatient'; 'the news that a group of physicists in England': James Conant, *My Several Lives: Memoirs of a Social Inventor*, Harper & Row, New York, 1970, pp. 278–80.

221 'the first I had heard about even the remote possibility of a bomb': Conant, *My Several Lives*, p. 277.

221 'practicable and likely to lead to decisive results': 'MAUD Committee Report', <www.atomicheritage.org/key-documents/maud-committee-report>, accessed 20 November 2021.

221 'they were not interested': Robyn Williams, 'Sir Mark Oliphant', *Life and Times*, 1985, radio program, ABC Radio National, <www.abc.net.au/rn/legacy/programs/lifeandtimes/stories/2009/2588019.htm>, accessed 11 November 2021.

222 'the all-out advocates of a head-on attack': as cited in Rhodes, *The Making of the Atomic Bomb*, p. 372.

222 'Professor Fermi was non-committal about the fast neutron bomb': as cited in Margaret Gowing, *Britain and Atomic Energy, 1939–1945*, Macmillan, London, 1964, p. 84, fn 2; Rhodes, *The Making of the Atomic Bomb*, p. 372

223 'Oliphant's story should be given serious consideration': as cited in Rhodes, *The Making of the Atomic Bomb*, p. 374.

224 Bush concerned about the Uranium Committee's lack of progress: Richard G Hewlett & Oscar E Anderson, Jr. *History of the United States Atomic Energy Commission*, vol. 1, *The New World, 1939–1946*, Pennsylvania State University Press, University Park, 1962, pp. 36–41.

224 'guess ... that Bush was secretly delighted': Stanley Goldberg, 'Inventing a climate of opinion: Vannevar Bush and the decision to build the bomb', *ISIS*, vol. 83, no. 3, September 1992, pp. 429–52, at p. 444.

225 'crushing disappointments and reverses': as cited in William Manchester & Paul Reid, *The Last Lion*, vol. 3, *Winston Spencer Churchill – Defender of the Realm, 1940–1965*, Little, Brown & Company, New York, 2012, p. 394.
226 'Enough evidence, we consider, has now been assembled': EP Abraham, E Chain, CM Fletcher, HW Florey, AD Gardner, NG Heatley & MA Jennings, 'Further observations on penicillin', *Lancet*, vol. 238, 1941, pp. 177–89, at p. 188.
227 finding work with the Moyer increasingly difficult: Gwyn Macfarlane, *Howard Florey: The Making of a Great Scientist*, Oxford University Press, Oxford, 1979, p. 342.
228 'too much planned for the Fermentation Division': Marc Landas, *Cold War Resistance: The International Struggle over Antibiotics*, Potomac Books, Lincoln, 2020, p. 43.
229 Lawrence the 'livewire': Hughes, *Australian Lives*, p. 128.
229 'They were as alike as two peas in a pod': as cited in Davis, *Lawrence and Oppenheimer*, p. 111.
229 'tolerate laziness or indifference'; 'at once noticed'; 'profound': 'Mark Oliphant, 'The two Ernests – Part 1', *Physics Today*, vol. 19, no. 9, 1966, pp. 33–49, at p. 40.
229 'The boy wonder of American science': in Brian VanDeMark, *Pandora's Keepers: Nine Men and the Atomic Bomb*, Back Bay Books, New York, 2005, p. 54.
230 Oliphant recounted his meetings with Briggs, Conant, Bush and Fermi: The sequence of Oliphant's meetings is disputed: Compare, for example, Cockburn & Ellyard's discussion, in Stewart Cockburn & David Ellyard, *Oliphant: The Life and Times of Sir Mark Oliphant*, Axiom Books, Adelaide, 1981, pp. 104–106; and Rhodes, *The Making of the Atomic Bomb*, pp. 372–4. I have been guided by Cockburn & Ellyard's more focused discussion of Oliphant's odyssey.
232 Oppenheimer tried to poison Patrick Blackett: Darren Holden, 'The indiscretion of Mark Oliphant: How an Australian kick-started the American atomic bomb project', *Historical Records of Australian Science*, vol. 29, no. 1, 2018, pp. 28–35 at, p. 32.
232 'an indiscretion – an eminent English [sic] visitor': Stephane Groueff, 'J Robert Oppenheimer interviewed by Stephane Groueff', 1965, <www.manhattanprojectvoices.org/oral-histories/j-robert-oppenheimers-interview>, accessed 26 November 2021; see too Darren Holden, 'Mark Oliphant and the Invisible College of the Peaceful Atom', PhD thesis, University of Notre Dame Australia, 2019, pp. 85–9, <researchonline.nd.edu.au/cgi/viewcontent.cgi?article=1278&context=theses>, accessed 26 November 2021.

Sources

233 'Whichever nation is first to succeed': Mark Oliphant, 'Memorandum to J. Robert Oppenheimer', 25 September 1941, as cited in Holden, 'Mark Oliphant and the Invisible College of the Peaceful Atom', p. 87.

Chapter 10

236 turning middling drug companies into the giants of big pharma: William Rosen, *Miracle Cure: The Creation of Antibiotics and the Birth of Modern Medicine*, Penguin Books, New York, 2017, pp. 169–72.

236 'Fortunately ... Richards appreciated the potential importance of the drug': Irivin Stewart, *Organizing Scientific Research for War: The Administrative History of the Office of Scientific Research and Development*, Atlantic-Little Brown, Boston, 1948, p. 105.

237 'Conant was reluctant': Arthur H Compton, *Atomic Quest: A Personal Narrative*, Oxford University Press, London, 1956, p. 7.

238 'Conant ... had come to Chicago still sharing Bush's scepticism': Michael Hiltzik, 'The man who saved the Manhattan Project', *History News Network*, 20 September 2015, <historynewsnetwork.org/article/160528>, accessed 20 January 2022; see also Michael Hiltzik, *Big Science: Ernest Lawrence and the Invention that Launched the Military-Industrial Complex*, Simon & Schuster, New York, 2015, pp. 213–30.

238 'the start of the wartime atomic race'; 'We in the United States': Compton, *Atomic Quest*, p. 53.

238 'the man who saved the Manhattan Project': Hiltzik, 'The man who saved the Manhattan Project'.

239 'the entire uranium project should be put in wraps': Compton, *Atomic Quest*, p. 49.

239 Third National Academy meeting on 21 October: Richard G Hewlett & Oscar E Anderson, Jr. *History of the United States Atomic Energy Commission*, vol. 1, *The New World, 1939–1946*, Pennsylvania State University Press, University Park, 1962, pp. 46–7.

241 finding the Frisch–Peierls Memorandum in a cornflake box: Lorna Arnold, 'The history of nuclear weapons: The Frisch–Peierls memorandum on the possible construction of atomic bombs of February 1940', *Cold War History*, vol. 3, no. 3, 2003, pp. 111–26, at pp. 112–4.

242 Developing the atomic bomb became an all-American enterprise: Neither the first comprehensive American report of the 'Atomic Bomb Project' published just days after the bombing of Hiroshima, nor the official history published in 1962 mention the Frisch–Peierls memorandum. See Henry De Wolf Smyth, *Atomic Energy for Military Purposes*, Maple Press, York, Pa.,

1945; and Hewlett & Anderson, *The New World*. Two good accounts of the
British contribution to the development of the atomic bomb are Sabina Lee,
'Birmingham – London – Los Alamos – Hiroshima: Britain and the Atomic
Bomb', *Midland History*, vol. 21, no. 1, 2002, pp. 146–64; and Ferenc Morton
Szasz, *British Scientists and the Manhattan Project: The Los Alamos Years*,
St Martin's Press, New York, 1992.

242 'arsenal of democracy': Arthur Herman, *Freedom's Forge: How American Business Produced Victory in World War II*, Random House, New York, 2013, pp. 128–9.

242 'I cannot escape the feeling that without active and continuing British interest'; British contribution 'helpful': Leslie R Groves, *Now It Can Be Told: The Story of the Manhattan Project*, Andre Deutsch, London, 1963, p. 408.

242 Oliphant liked Groves: Clarence Larson, *Pioneers of Science and Technology: Interview with Mark Oliphant*, Washington DC, 11 March 1988, <www.youtube.com/watch?v=z0QaXvLsR9A>, accessed 10 November 2021.

242 'moderate attainments'; 'in no sense vital': as cited in Ferenc Morton Szasz, *British Scientists and the Manhattan Project: The Los Alamos Years*, St Martin's Press, New York, 1992, p. xv.

243 This is ludicrous: Sabine Lee, '"Crucial? Helpful? Practically nil?" Reality and perception of Britain's contribution to the development of nuclear weapons during the Second World War', *Diplomacy and Statecraft*, vol. 33, no. 1, 6 May 2022, pp. 19–40.

243 'immediate catalyst'; 'British science ... had won admiration': McGeorge Bundy, *Danger and Survival: Choices about the Bomb in the First Fifty Years*, Vintage Books, New York, 1990, p. 48.

243 'controlling element': as cited in Bundy, *Danger and Survival*, p. 49.

243 'I, personally, was responsible': as cited in Stewart Cockburn & David Ellyard, *Oliphant: The Life and Times of Sir Mark Oliphant*, Axiom Books, Adelaide, 1981, p. 109.

243 'Who really invented the atom bomb': as cited in Christoph Laucht, *Elemental Germans: Klaus Fuchs, Rudolf Peierls and the Making of British Nuclear Culture, 1939–59*, Palgrave Macmillan, London, 2012, p. 39.

244 'fascinating and impressive': Arnold, 'The history of nuclear weapons', at p. 111.

245 atomic bomb was 'feasible': Ronald W Clark, *The Birth of the Bomb: The Untold Story of Britain's Part in the Weapon that Changed the World*, Phoenix House, London, 1961, p. 137.

245 'One good way to understand the force': Bundy, *Danger and Survival*, p. 25.

Sources

245 'compelled in some way or other to work for Hitler': Charles Weiner, 'Oral histories: Otto Frisch', Niels Bohr Library and Archives, American Institute of Physics, 3 May 1967, <www.aip.org/history-programs/niels-bohr-library/oral-histories/4616>, accessed 15 November 2021.

246 Oliphant's 'decisive' influence: see Gowing, *Britain and Atomic Energy*, pp. 117, 121. Hewlett & Anderson's *The New World* (at pp. 42–4) is more reticent but also acknowledges Oliphant's role.

246 'Oliphant convinced Lawrence': Richard Rhodes, *The Making of the Atomic Bomb*, Simon & Schuster, New York, 1986, p. 377.

247 'Oliphant came over from England': Spencer Weart & Gertrude Szilard (eds), *Leo Szilard: His Version of the Facts (Selected Recollections and Correspondence)*, MIT Press, Cambridge, Mass., 1978, vol. 2, p. 146.

247 'What [Roosevelt] did [in late 1939]'; 'He woke up America very effectively': Cindy Kelly, 'Lawrence Bartell interviewed by Cindy Kelly, Ann Arbor, Michigan', 9 May 2013, <www.manhattanprojectvoices.org/oral-histories/lawrence-bartells-interview>, accessed 29 November 2021.

Chapter 11

249 'salvation of the Allied cause': as cited in Jennet Conant, *Tuxedo Park: A Wall Street Tycoon and the Secret Palace of Science that Changed the Course of World War II*, Simon & Schuster, New York, 2003, p. 191.

249 'The only thing that ever really frightened me': Winston Churchill, *The Second World War*, vol. 2, *Their Finest Hour*, Houghton Mifflin, Boston, 1949, p. 598.

250 'pearl beyond price': Conant, *Tuxedo Park*, p. 182.

250 'Happy Time': Norman Fine, *Blind Bombing: How Microwave Radar Brought the Allies to D-Day and Victory in World War II*, University of Nebraska Press, Lincoln, 2019, p. 85.

250 'mile-long freight train': William Manchester & Paul Reid, *The Last Lion*, vol. 3, *Winston Spencer Churchill – Defender of the Realm, 1940–1965*, Little, Brown & Company, New York, 2012, p. 515.

252 Operation Drumbeat claimed 5000 lives: Jonathan Dimbleby, *The Battle of the Atlantic: How the Allies Won the War*, Penguin, London, 2016, p. 264.

252 'Everything we knew': as cited in Jeremy Bernstein, 'Profiles: I. I. Rabi', *New Yorker*, 20 October 1975, pp. 47–8, at p. 48.

253 'the greatest cooperative research establishment': as cited in TA Saad, 'The story of the MIT radiation laboratory', *IEEE Aerospace and Electronic Systems Magazine*, vol. 5, no. 10, 1990, pp. 46–51, at p. 47.

253 'like currants in a cake': EG Bowen, *Radar Days*, Hilger, Bristol, 1987, p. viii.

254 the War Cabinet considered cavity magnetron secret and valuable: DCT Bennett, *Pathfinder: A War Autobiography*, Goodall Publications Limited, London, 1988, p. 152.

254 the proximity fuze: a tiny radio transmitter: 'The proximity fuze the smallest radar' in Brown, See Louis Brown, *Technical and Military Imperatives: A Radar History of World War II*, Institute of Physics Publishing, Bristol, 1999, pp. 174–86.

254 'if it had not been for radar and the A-bomb': Vannevar Bush, 'Foreword', in James Phinney Baxter, *Scientists Against Time*, 2nd edn, MIT Press, Cambridge, Mass., 1968, p. x.

255 'a fear that the British had developed': Karl Doenitz, *Memoirs: Ten Years and Twenty Days*, Weidenfeld & Nicolson, London, 1959, p. 236.

255 effectively quadrupling the area 'visible' in flight: Henry Guerlac & Marie Boas, 'The radar war against the U-boat', *Military Affairs*, vol. 14, no. 2, Summer 1950, pp. 99–111, at p. 106.

255 ability to spot the needle in a haystack: John G Shannon & Paul M Moser, 'A history of US Navy periscope detection radar: Sensor design and development', Office of Naval Research, 2014, <apps.dtic.mil/dtic/tr/fulltext/u2/1003753.pdf%20accessed%201%20December%202021>, accessed 26 November 2021.

256 'pre-eminence in microwave radar'; 'In areas where air cover was strong': Doenitz, *Memoirs*, p. 252.

256 In 1942, U-boats sank more tons of shipping than the Allies managed to build: Guerlac & Boas, 'The radar war against the U-boat', p. 104.

257 'in the first twenty days of March 1943': as cited in AJP Taylor, *The Second World War: An Illustrated History* [1975] in *The Second World War and Its Aftermath*, Folio Society, London, 1998, p. 120.

257 One hundred and seven merchant ships went down: Guerlac & Boas, 'The radar war against the U-boat', pp. 104–5.

257 'the worst month of the war': Taylor, *The Second World War*, p. 120.

257 'effect'; 'instantaneous and dramatic': as cited in Dimbleby, *The Battle of the Atlantic*, p. 401.

258 'the centimetric radar set': as cited in Robert Buderi, *The Invention that Changed the World: The Story of Radar from War to Peace*, Little, Brown & Company, London, 1997, p. 169.

258 'Radar, and particularly radar location': Doenitz, *Memoirs*, p. 341.

258 Hitler agreed, conceding that radar: Stewart Cockburn & David Ellyard, *Oliphant: The Life and Times of Sir Mark Oliphant*, Axiom Books, Adelaide, 1981, p. 89.

Sources

259 'military mindset': Buderi, *The Invention that Changed the World*, p. 202.
259 Corporate rivalry as well as licensing and patent issues: John Cornwall, *Hitler's Scientists: Science, War, and the Devil's Pact*, Penguin, New York, 2004, p. 265.
259 'Battle of the Beams': see, RV Jones, *Most Secret War: British Scientific Intelligence 1939–1945*, Coronet, London, 1979, pp. 134–44.
260 'We must frankly admit': as cited in Cornwall, *Hitler's Scientists*, p. 280.
260 Japan was not far behind in radar technology: Buderi, *The Invention that Changed the World*, pp. 240–242.
260 best suited for installation on the Japanese home islands: Buderi, *The Invention that Changed the World*, pp. 242–44.
263 'almost mad with terror': Max Hastings, *Nemesis: The Battle for Japan, 1944–45*, Harper Collins, London, 2007, p. 324.
263 'were iron safes standing forlorn amid the ashes': Hastings, *The Battle for Japan*, p. 327.
264 'the most valuable English scientific innovation': CP Snow, *The Physicists*, Macmillan, London, 1981, p. 105.
264 'salvation of the Allied cause': as cited in Conant, *Tuxedo Park*, p. 191.
264 'This revolutionary discovery': Baxter, *Scientists Against Time*, Little, p. 142.
264 'the breakthrough': as cited in Raphael Chayim Rosen, 'Under the radar physics, engineering, and the distortion of a World War Two legacy', honors thesis, Harvard University, Cambridge, Mass., March 2006, fn 82.
265 'the resonant cavity magnetron': Fine, *Blind Bombing*, p. 2.
265 'If the British victory at the Battle of El-Alamein': Dimbleby, *The Battle of the Atlantic*, p. 423.
265 'All through the war': as cited in Dimbleby, *The Battle of the Atlantic*, p. 22.

Chapter 12

266 might have convinced American drug companies: AN Richards, 'Production of penicillin in the United States (1941–1946)', *Nature*, no. 4918, 1 February 1964, pp. 441–5.
266 'Takes eleven acres of mold': Harry C Butcher, *My Three Years with Eisenhower*, Simon & Schuster, New York, 1946, p. 12.
266 increased penicillin production: Richards, 'Production of penicillin'.
267 one of out the thousand: Robert Gaynes, 'The discovery of penicillin – New insights after more than 75 years of clinical use', *Emerging Infectious Diseases*, vol. 23, no. 5, May 2017, pp. 849–53, at p. 850.

267 Salvation came from a woman named Mary Hunt; Hunt's role is contested: Ronald Kotulak, 'Sorry Mary, you're no hero after all: Real penicillin star is unknown housewife', *The Dispatch (Moline, Illinois)*, 10 October 1976, <clickamericana.com/topics/health-medicine/how-penicillin-was-discovered>, accessed 8 December 2021.

269 as Ernst Chain had sadly predicted: Eric Lax, *The Mold in Dr. Florey's Coat: The Story of the Penicillin Miracle*, Henry Holt Company, New York, 2004, pp. 162–4.

269 new techniques developed in America: William Rosen, *Miracle Cure: The Creation of Antibiotics and the Birth of Modern Medicine*, Penguin Books, New York, 2017, pp. 156–7.

270 Ethel Florey was put in charge of the second clinical trial: HW Florey et al., *Antibiotics*, vol. 2, Oxford University Press, London, 1949, pp. 651–6.

270 'My wife is doing the clinical work': as cited in Gwyn Macfarlane, *Howard Florey: The Making of a Great Scientist*, Oxford University Press, Oxford, 1979, p. 346.

270 'a corpse-reviver': as cited in Macfarlane, *Howard Florey*, p. 345.

272 'The forest of uncertainty': Andrew Hargadon, 'It's not about the idea', *TEDxSacramento*, 12 December 2012, <www.youtube.com/watch?v=wtsdkq97AMw>, accessed 8 December 2021.

272 'Notwithstanding claims made years later': Trevor I Williams, *Howard Florey: Penicillin and After*, Oxford University Press, Oxford, 1984, p. 71.

272 'a 2000-year history of an idea': Hargadon, 'It's not about the idea'.

273 'there is, for me, no doubt': as cited in Macfarlane, *Howard Florey*, p. 346.

274 'Sir, In the leading article on penicillin': as cited in Macfarlane, *Howard Florey*, p. 349.

275 'limited skills in diplomacy': Edward Abraham, 'Oxford, Howard Florey and World War II' in Carol L Moberg & Zanvil A Cohn (eds), *Launching the Antibiotic Era: Personal Accounts of the Discovery and Use of the First Antibiotics*, Rockefeller University Press, New York, 1990, pp. 19–30, p. 30.

275 'In many ways … it was as much my father's fault': Max Blythe, 'Professor Charles du Vé Florey in interview with Dr Max Blythe, Edinburgh, 17 February 1998', Royal College of Physicians and Oxford Brookes University Medical Sciences Video Archive MSVA 179, <radar.brookes.ac.uk/radar/file/7ea51825-d26c-4575-8973-985d58aec234/1/Florey%2CC.pdf>, accessed 13 January 2022.

275 When reporters turned up one day at the Dunn School: Lax, *The Mold in Dr. Florey's Coat*, p. 212.

275 'unscrupulous campaign': as cited in Macfarlane, *Howard Florey*, p. 354.

Sources

276 'Gentlemen, the Government will provide': as cited in Lennard Bickel, *Rise up to Life: A Biography of Howard Walter Florey Who Made Penicillin and Gave It to the World*, Angus & Robertson, London, 1972, p. 170.

277 'The discovery was old science': Bill Bynum, 'Shedding new light on the story of penicillin', *Lancet*, vol. 369, 16 June 2007, pp. 1991–2, at p. 1991.

277 'Howard Florey can ... be regarded as a model'; 'What happened then is a paradigm': Peter Doherty, 'Time to champion our tall poppies', *Australian Quarterly*, vol. 70, no. 6, Nov–Dec 1998, pp. 54–6, at p. 56.

278 'Florey was not a profound visionary': Henry Harris, 'Howard Florey and the development of penicillin', *Notes and Records: The Royal Society Journal of the History of Science*, vol. 53, no. 2, 1999, pp. 243–52, at p. 252.

278 'commitment in the face of uncertainty'; 'the defining characteristic of innovation': Andrew Hargadon, 'It's not about the idea'.

280 'This valuable drug must on no account be wasted': cited in Jonathan Liebenau, 'The British success with penicillin', *Social Studies of Science*, vol. 17, no. 1, 1987, pp. 69–86, at p. 79.

280 'The goal ... to make penicillin so cheaply': cited in Williams, *Howard Florey*, p. 182.

280 the price falling from $20: Robert Bud, *Penicillin: Triumph and Tragedy*, Oxford University Press, Oxford, 2007, p. 53.

280 'Penicillin alone was worth more than an Army corps': as cited in James Phinney Baxter, *Scientists Against Time*, 2nd edn, MIT Press, Cambridge, Mass., 1968, p. 106.

280 'We've snatched them right out the grave': as cited in Rick Atkinson, *The Day of Battle: The War in Sicily and Italy, 1943–1944*, Henry Holt, New York, 2007, p. 493.

281 only the atomic bomb had higher priority: Gerald Jonas, *The Circuit Riders: Rockefeller Money and the Rise of Modern Science*, WW Norton, New York, 1989, p. 305.

281 Churchill became sick on the trip; recovered using standard sulpha drugs: Bickel, *Rise up to Life*, p. 211.

282 Florey was aware of German interest in penicillin: Gilbert Sharma & Jonathan Reinarz, 'Allied intelligence reports on wartime German penicillin research and production', *Historical Studies in the Physical and Biological Sciences*, vol. 32, no. 2, 2002, pp. 347–61, at p. 348.

282 'During the past 10 years I have sent out a very large number of cultures': as cited in Gaynes, 'The discovery of penicillin', p. 851.

282 'the German guard was kept drunk': Gaynes, 'The discovery of penicillin', p. 851.

283 'a screwball only interested in money': cited in Milton Wainwright, 'Hitler's penicillin', *Perspectives in Biology and Medicine*, vol. 47, no. 2, 2004, pp. 189–98, at p. 194.

283 'arguably did save Hitler's life': Wainwright, 'Hitler's penicillin', p. 197.

284 'After fighting disease and battle infections': Dean Anderson, *Praise the Lord and Pass the Penicillin: Memoir of a Combat Medic in the Pacific in World War II*, McFarland, Jefferson, NC, 2003, p. 1.

285 7-year-old Peter Harrison: Ameneh Khatami, Philip N Britton, Glendon Farrow, Megan Phelps, Alyson Kakakios, 'Meningitis and the military: the remarkable story of the first use of penicillin in Australia', *MJA*, vol. 213, no. 11, 14 December 2020, pp. 508–10.

285 'The time has come': as cited in Bickel, *Rise up to Life*, p. 232.

286 'the spectacle of everybody': as cited in Lax, *The Mold in Dr. Florey's Coat*, p. 232.

287 'In time, even the public will realize that': as cited in Lax, *The Mold in Dr. Florey's Coat*, p. 235.

288 'the considered opinion of a large group'; 'Florey, Chain and Fleming': as cited in Lax, *The Mold in Dr. Florey's Coat*, p. 238.

289 'PROFOUNDLY DISTURBED OVER': as cited in Lax, *The Mold in Dr. Florey's Coat*, p. 239.

289 Nobel prize 'for the discovery of penicillin': as cited in Lax, *The Mold in Dr. Florey's Coat*, p. 247.

Chapter 13

291 'I feel quite sure that in your hands': as cited in Stewart Cockburn & David Ellyard, *Oliphant: The Life and Times of Sir Mark Oliphant*, Axiom Books, Adelaide, 1981, p. 108.

292 'one of the most effective scientific committees': Margaret Gowing, *Britain and Atomic Energy, 1939–1945*, Macmillan, London, 1964, p. 80.

292 'I can see no reason': as cited in Cockburn & Ellyard, *Oliphant*, pp. 108–9.

294 'It appears desirable that we should soon correspond': as cited in Diana Preston, *Before the Fall-Out: The Human Chain Reaction from Marie Curie to Hiroshima*, Corgi Books, London, 2006, p. 266.

295 'We must, however, face the fact': as cited in Jim Baggott, *Atomic: The First War of Physics and the Secret History of the Atom Bomb, 1939–1949*, Icon, London, 2015, p. 142.

298 Oliphant claimed that he was also called to the White House: Michael Wilson, personal communication, June 2021.

Sources

300 'Lawrence–Oliphant electromagnetic method'; 'racetracks': Richard Cohen, *Chasing the Sun: The Epic Story of the Star that Gives Us Life*, Simon & Schuster, London, 2011, p. 215.

301 'A hatred of the Hitler regime': as cited in Robin Hughes, *Australian Lives: Stories of Twentieth Century Australians*, Angus & Robertson, Sydney, 1996, p. 133.

301 'We must make use of physics for warfare': as cited in Cohen, *Chasing the Sun*, p. 199.

302 uranium from the mines of Joachimsthal: Giles MacDonogh, *After the Reich: From the Liberation of Vienna to the Berlin Airlift*, John Murray, London, 2008, p. 213.

302 Heisenberg's protestations can safely be dismissed: Baggott, *Atomic*, pp. 351–4.

302 'we saw an open road ahead of us, leading to the atomic bomb': as cited in Preston, *Before the fallout*, p. 266.

302 the Nazi hierarchy did not pursue it: Stanley Goldberg, 'Inventing a climate of opinion: Vannevar Bush and the decision to build the bomb', *ISIS*, vol. 83, no. 3, September 1992, pp. 429–52, pp. 450–2.

303 'I would definitely have considered making atomic bombs for Hitler a crime': Sven Felix Kellerhof, 'Hitler's physicists and the bomb', 1 February 2015, <www.ozy.com/true-and-stories/hitlers-physicists-and-the-bomb/37426/>, accessed 28 November 2021.

303 'It may interest you': cited in Goldberg, 'Inventing a climate of opinion', p. 435.

304 'the German nuclear scientists have stopped publishing!': as cited Cohen, *Chasing the Sun*, p. 213 fn.

305 Japan also entertained the possibility of an atomic bomb: 'Japanese atomic bomb project', Atomic Heritage Foundation, <www.atomicheritage.org/history/japanese-atomic-bomb-project>, accessed 28 November 2021.

305 'And then, without a sound, the sun was shining': Otto Frisch, *What Little I Remember*, Cambridge University Press, New York, 1979, p. 164.

306 'We were struck with awe.': Rudolf Peierls, *Bird of Passage: Recollections of a Physicist*, Princeton University Press, Princeton, 1985 p. 202.

307 'an elongated trash can with fins': Richard Rhodes, *The Making of the Atomic Bomb*, Simon & Schuster, New York, 1986, p. 701.

308 'Our orders were for a visual bombing run': Paul W Tibbets, *Return of the Enola Gay*, Enola Gay Remembered, New Hope, Pa., 1998, pp. 224–5.

309 the Manhattan Project at $2.2 billion: Richard G Hewlett & Oscar E Anderson, Jr. *The New World, 1939–1946*, vol. 1, *A History of the Atomic Energy Commission*, Pennsylvania State University Press, University Park, 1962, pp. 723–4.

309 B-29 bomber (developed at a cost of $3 billion): Walter J Boyne, 'The B-29's Battle of Kansas', *Air Force Magazine*, 1 February 2012, <www.airforcemag.com/article/0212b29/>, accessed 14 December 2021.

309 microwave radar costing about $3 billion: James Phinney Baxter, *Scientists Against Time*, 2nd edn, MIT Press, Cambridge, Mass., 1968, p. 142.

309 *Time* magazine, which was planning to put radar on the cover: Reynolds, *The Long Shadow*, p. 301; *Time*, vol. XLVI, no. 820, August 1945, <content.time.com/time/magazine/0,9263,7601450820,00.html>, accessed 8 December 2021.

310 Oliphant himself did not leave a contemporary reflection: Cockburn & Ellyard, *Oliphant*, p. 123.

310 'unease, indeed nausea'; 'seemed rather ghoulish': Frisch, *What Little I Remember*, p. 176.

310 was the nuclear double whammy at all necessary?: For a survey of the arguments about the Japanese surrender see 'Debate over the Japanese surrender', Atomic Heritage Foundation, <www.atomicheritage.org/history/debate-over-japanese-surrender>, accessed 13 January 2022.

311 primarily a warning to the Soviets: see for example Kathryn Keeble, 'General Groves's "inevitable war with Russia": Joseph Rotblat's and Mark Oliphant's existential crises', *Journal of Australian Studies*, 15 October 2021, <doi.org/10.1080/14443058.2021.1987293>, accessed 20 January 2022.

312 'we broke down and cried with relief and joy': Paul Fussell, 'Thank God for the atom bomb', in *Thank God for the Atom Bomb and Other Essays*, Summit Books, New York, 1988, pp. 13–37, at p. 28.

312 'There's a lot of talk now': Shelby Foote, 'The art of fiction no. 158', *Paris Review*, Issue 151, Summer 1999, <www.theparisreview.org/interviews/931/the-art-of-fiction-no-158-shelby-foote>, accessed 13 January 2022.

313 Japan's plan for the defence of homeland: Thomas B Allen & Norman Polmar, *Codename Downfall: The Secret Plan to Invade Japan*, Headline Book Publishing, London, 1995, pp. 259ff.

313 predicted up to a million American casualties: Allen & Polmar, *Codename Downfall*, pp. 238–46, 248–52, 356–7.

314 American military planners expected Japanese resistance to end in November 1946: Allen & Polmar, *Codename Downfall*, p. 255.

314 'endur[e] the unendurable'; 'The enemy has begun to employ a new and most cruel bomb'; 'jewelled sound': Edwin P Hoyt, *Japan's War: The Great Pacific Conflict 1853–1952*, Arrow Books, London, 1989, Appendix D, pp. 437–8.

Sources

315 the atomic bomb ended the Second World War, but microwave radar won it: Daniel Kevles, *The Physicists*, Alfred A Knopf, New York, 1978, p. 308.

315 'had a more decisive effect on the outcome': AP Rowe, *One Story of Radar*, Cambridge University Press, Cambridge, 1948, p. 35.

316 'I'm very serious about this war': as cited in Rhodes, *The Making of the Atomic Bomb*, p. 452.

316 'World War II was the first war in human history': Vannevar Bush, 'Foreword', in Baxter, *Scientists Against Time*, p. ix.

Chapter 14

317 'You know, if you leave this country'; 'It'll be the end of your research career!'; 'You know what you'll find when you get there?': as cited in Robin Hughes, *Australian Lives: Stories of Twentieth Century Australians*, Angus & Robertson, Sydney, 1996, p. 140.

319 'downright dishonesty': as cited in Eric Lax, *The Mold in Dr. Florey's Coat: The Story of the Penicillin Miracle*, Henry Holt & Company, New York, 2004, p. 250.

319 'In a time when annihilation and destruction …': G Liljestrand, 'Award ceremony speech: Presentation Speech by Professor G. Liljestrand, member of the Staff of Professors of the Royal Caroline Institute, on December 10, 1945', < www.nobelprize.org/prizes/medicine/1945/ceremony-speech/>, accessed 24 November 2021.

319 'Penicillin was made available …': as cited in Ernst B Chain, 'Banquet speech: Ernst B. Chain's speech at the Nobel Banquet in Stockholm, December 10, 1945', <www.nobelprize.org/prizes/medicine/1945/chain/speech/>, accessed 24 November 2021.

320 'You have become the most famous doctors in the whole world': as cited in Ernst B Chain, 'Banquet speech'.

321 part war hero and part war criminal: John Larkin, 'Genius and the genie: Sir Mark Oliphant on the atomic bomb, Interview by John Larkin', *Age*, 8 August 1985, p. 11.

321 'eminent scientists of the United Kingdom, Canada, Australia and Russia': Stewart Cockburn & David Ellyard, *Oliphant: The Life and Times of Sir Mark Oliphant*, Axiom Books, Adelaide, 1981, p. 196.

322 'Doctor Marcus Laurence Elwin Oliphant, Australian citizen': Cockburn & Ellyard, *Oliphant*, p. 196.

323 Nancy Wake Medal of Freedom: Peter FitzSimons, *Nancy Wake: A Biography of Our Greatest War Heroine*, Harper Collins Publishers, Sydney, 2001, p. 284.

323 Synchrotron, revolutionary technology would dominate research: David Ellyard, 'Mark Oliphant FRS and the Birmingham proton synchrotron', PhD thesis, University of New South Wales, December 2011, <unsworks.unsw.edu.au/fapi/datastream/unsworks:10945/SOURCE01?view=true>, accessed 25 November 2021.

323 Oliphant first proposed a synchrotron: EJN Wilson, 'Fifty Years of synchrotron', 5th European Particle Accelerator Conference, Sitges, Barcelona, 10–14 June 1996, <accelconf.web.cern.ch/e96/papers/orals/frx04a.pdf>, accessed 25 November 2021.

324 Oliphant's insight might have been worthy of the Nobel prize: Cockburn & Ellyard, *Oliphant*, p. 138.

324 'We wanted Florey, and we wanted Oliphant': as cited in Lennard Bickel, *Rise up to Life: A Biography of Howard Walter Florey Who Made Penicillin and Gave It to the World*, Angus & Robertson, London, 1972, pp. 262–3.

325 'riding the breakers and browning himself': WK Hancock, *Country and Calling*, Faber & Faber, London, 1954, p. 239.

326 'It may be ... that for science generally': Mark Oliphant, 'Some recollections of Lord Florey', *ANU News*, April 1969, pp. 8–9, at p. 9.

326 'was at his spell-binding best'; 'The impact on Chifley': as cited in SC Foster & Margaret M Varghese, *The Making of the Australian National University, 1946–1996*, Allen & Unwin, Sydney, 1996, p. 21.

327 Oliphant knew both May and Fuchs: Roland Perry, 'The professor and the atom bomb', *Age*, 22 July 2000, News Extra, p. 2.

327 Oliphant was denied a visa: Cockburn & Ellyard, *Oliphant*, pp. 182–201.

327 Chain also denied a visa: Ronald W Clark, *The Life of Ernst Chain: Penicillin and Beyond*, Weidenfeld & Nicolson, London, 1985, pp. 127–8.

328 'I said I would come because I was ... a very loyal Australian': Hughes, *Australian Lives*, pp. 139–40.

328 Florey as President of the Royal Society: Trevor I Williams, *Howard Florey: Penicillin and After*, Oxford University Press, Oxford, 1984, pp. 320–45.

329 'It is good to know that some real science is to be infiltrated into the House of Lords': Mark Oliphant, Letter to Howard Florey, 5 January 1965, Papers of Howard Walter Florey, Royal Society, Series 225, Item 235.

330 'I could quite readily dispense': as cited in Gwyn Macfarlane, *Howard Florey: The Making of a Great Scientist*, Oxford University Press, Oxford, 1979, p. 370.

330 'he was the best and soundest builder': Macfarlane, *Howard Florey*, p. 262.

331 'My dear Oliphant': Howard Florey, Letter to Mark Oliphant, 29 October 1966, Sir Marcus Laurence Oliphant Papers, University of Adelaide Library Rare Books and Special Collection, MSS 92 O4775p, Series 26.

Sources

331 Florey's health suffered: Dr Godfrey Fowler OBE, Florey's physician, interview with the author, May 2018.

332 'I carried on and I did ruin my research career': Mark Oliphant, Letter to the Vice Chancellor, Australian National University, 16 September 1991, Sir Marcus Laurence Oliphant Papers, University of Adelaide Library Rare Books and Special Collection, MSS 92 O4775p, Series 26.

333 'the White Oliphant': Conrad Roberts, 'The White Oliphant?', *Bulletin*, 25 January 1961, pp. 12–15.

333 'For myself I regret it': Hughes, *Australian Lives*, p. 142.

333 'belligerent pacifist': as cited in Sue Rabbitt Roff, 'Was Sir Mark Oliphant Australia's—and Britain's—J. Robert Oppenheimer?', *Meanjin Quarterly*, 22 January 2019, <meanjin.com.au/uncategorised/was-sir-mark-oliphant-australias-and-britains-j-robert-oppenheimer/>, accessed 14 December 2021.

334 Oliphant soured on nuclear weapons: Darren Holden, '"On the Oliphant deign, now sound the blast"; How Mark Oliphant secretly warned of America's post-war intentions of an atomic monopoly', *Historical Records of Australian Science*, vol. 29, no. 2, 2018, pp. 130–7; Anna Binnie, 'Oliphant, the father of atomic energy', *Journal & Proceedings of the Royal Society of New South Wales*, vol. 139, no. 419, 2006, pp. 11–22.

334 Oliphant became a convert to renewable energy: Monica Oliphant, interview with author, May 2021.

334 'quite clear to me that the university': Clarence Larson, *Pioneers of Science and Technology: Interview with Mark Oliphant*, Washington DC, 11 March 1988, <www.youtube.com/watch?v=z0QaXvLsR9A>, accessed 10 November 2021.

334 'approach may differ considerably from the accepted norm': as cited in Cockburn & Ellyard, *Oliphant*, p. 284.

335 'a lifelong activist': Michael Wilson, interview with author, June 2021.

335 Oliphant campaign for voluntary euthanasia: Trevor Gill, 'Sir Mark: I would prefer to die …', *Sunday Mail*, 21 July 1985, p. 47.

Afterword

337 'One can think': as cited in Robert Bud, *Penicillin: Triumph and Tragedy*, Oxford University Press, Oxford, 2007, p. 3.

337 estimated 200 million lives saved by penicillin: 'One discovery that changed the World', *Florey 120 Anniversary*, 2018, <health.adelaide.edu.au/florey120anniversary/one-discovery-that-changed-the-world>, accessed 25 November 2021.

337 an estimated 66 million men, women and children suffering from tuberculosis: 'Tuberculosis: Key facts', World Health Organization, 14 October 2021, <www.who.int/news-room/fact-sheets/detail/tuberculosis>, accessed 25 November 2021.

337 23 years added to our life expectancy: Matthew Hutchings, Andrew W Truman & Barrie Wilkinson, 'Antibiotics: Past, present and future', *Current Opinion in Microbiology*, vol. 51, October 2019, pp. 72–80.

338 one of the few recordings of Florey's voice: Hazel de Berg, 'Interview notes, Interview with Howard Florey, Tape 220–21', 5 April 1967; Barry York, 'Howard Florey and the development of penicillin', *National Library of Australia News*, vol. XI, no. 12, September 2001, pp. 18–20, at pp. 19–20.

338 'People sometimes think that': Hazel de Berg, 'Howard Walter Florey interviewed by Hazel de Berg', 5 April 1967, *Hazel de Berg Collection*, National Library of Australia Oral History Transcript, <nla.gov.au/nla.obj-220872612>, accessed 13 January 2022.

338 'Florey was totally averse': Lennard Bickel, *Rise up to Life: A Biography of Howard Walter Florey Who Made Penicillin and Gave It to the World*, Angus & Robertson, London, 1972, p. 294.

340 'The atomic bomb was a sideshow': Robert Buderi, *The Invention that Changed the World: The Story of Radar from War to Peace*, Little, Brown & Company, London, 1997, back cover encomium.

340 'one scientific weapon [that] meant the difference between victory and defeat': David E Fisher, *A Race on the Edge of Time: Radar – The Decisive Weapon of World War II*, McGraw Hill, New York, 1989, p. xi.

341 'Pulses into Ploughshares': Robert Watson-Watt, *Three Steps to Victory: A Personal Account By Radar's Greatest Pioneer*, Odhams Press, London, 1957, p. 442.

341 *Titanic* would have safely reached New York: Sharon Gaudin, 'Titanic was high-tech marvel of its time', *Computerworld*, 14 April 2012, <www.computerworld.com/article/2503092/titanic-was-high-tech-marvel-of-its-time.html?page=2>, accessed 25 November 2021.

341 one billion microwave ovens: Evan Ackerman, 'A brief history of the microwave oven where the "radar" in Raytheon's Radarange came from', *IEEE Spectrum*, 30 September 2016, <spectrum.ieee.org/a-brief-history-of-the-microwave-oven>, accessed 25 November 2021.

342 'Oliphant's heroic efforts': 'Manhattan Project history: Britain's early input – 1940–41', *Atomic Heritage Foundation*, <www.atomicheritage.org/history/britains-early-input-1940-41>, accessed 25 November 2021.

Sources

343 'There is only one question': William Faulkner, speech at the Nobel Banquet, 10 December 1950, <www.nobelprize.org/prizes/literature/1949/faulkner/speech/>, accessed 13 January 2022.

343 'nothing so menacing to our civilization'; 'poignant to look at youth': as cited in AN Wilson, *After the Victorians: 1901–1953*, Hutchinson, London, 2005, pp. 462–3.

343 'It seems inescapable that what has really made the difference': John Lewis Gaddis, *The Long Peace: Inquiries into the History of the Cold War*, Oxford University Press, New York, 1987, p. 230.

343 Deterrence has been a contentious view: With the end of the Cold War and emergence of new nuclear states without practice safeguards (North Korea and possibly Iran), increasing possibilities of regional nuclear arms races and nuclear terrorism, and even that the wholesale destruction of cities wrought by nuclear weapons is militarily ineffective, the theory of nuclear deterrence is now routinely challenged. For a seminal critique see: Ward Wilson, 'The myth of nuclear deterrence', *Nonproliferation Review*, vol. 15, no. 3, November 2008, pp. 421–39.

344 18 percent of global electricity: Peter Fairley, 'Why don't we have more nuclear power?', *MIT Technology Review*, 28 May 2015, <www.technologyreview.com/2015/05/28/167951/why-dont-we-have-more-nuclear-power/>, accessed 25 November 2021.

344 Proponents of nuclear power: Michael Shellenberger, *Apocalypse Never: Why Environmental Alarmism Harms Us All*, Harper Collins, New York, 2020, pp. 151–5; Chapter 15, Richard Rhodes, *Energy: A Human History*, Simon & Schuster, New York, 2018, Chapter 20.

344 Not everyone is convinced by nuclear energy: see for example Mark Diesendorf, 'Is nuclear energy a possible solution to global warming?', Institute of Environmental Studies, University of New South Wales, <www.ceem.unsw.edu.au/sites/default/files/uploads/publications/NukesSocialAlternativesMD.pdf>, accessed 13 January 2022; Oliphant turned against nuclear energy: Stewart Cockburn & David Ellyard, *Oliphant: The Life and Times of Sir Mark Oliphant*, Axiom Books, Adelaide, 1981, pp. 130–1.

344 'I still think ... that science can save mankind': as cited in 'Oliphant remembered as a truly great Australian', *7:30 Report*, 17 July 2000, television program, ABC TV.

345 'Florey ... knew not only his science': 'People of science with Brian Cox: Dame Sally Davies – Alexander Fleming and Howard Florey', <royalsociety.org/about-us/programmes/people-of-science/sally-davies-fleming-florey/>, accessed 25 November 2021.

347 'Great Australian Emptiness': Patrick White, 'Prodigal Son', in Imre Salusinszky (ed.), *The Oxford Book of Australian Essays*, Oxford University Press, Melbourne, 1997, pp. 125–8, at p. 126.
347 'disease of the Australian mind': AA Phillips, 'The cultural cringe', *Meanjin*, vol. 9, no. 4, 1950, pp. 299–302.
348 'stifling intellectual torpor': 'Brilliant creatures: Germaine, Clive, Barry & Bob', ABC TV, 23 September 2014, <iview.abc.net.au/show/brilliant-creatures-germaine-clive-barry-and-bob>, accessed 25 November 2021.
348 'for an Australian, everywhere was fairyland': Peter Craven, 'Peter Conrad's *How the World Was Won* explores the romance of America', *Weekend Australian*, 'Review' section, 22–23 November 2014, p. 22.
348 'to the outsider ... Australia is a very conservative country': de Berg, 'Howard Walter Florey interviewed by Hazel de Berg'.
348 'Can't we do anything ourselves as Australians?': CMH Clarke, *A History of Australia*, vol. 6, Melbourne University Press, Melbourne, 1987, p. vii.
348 'much more happy in our skin': as cited in Andrew Clark, 'Young, free and confident', *Australian Financial Review*, 24 January 2015, p. 18.
349 13 Australians have been awarded a Nobel prize: 'Nobel Australians', Australian Academy of Science, <www.science.org.au/education/history-australian-science/nobel-australians>, accessed 25 November 2021. I have excluded from the original list of 15 the Queensland-born Prokhorov, who moved to the Soviet Union when he was 7, and Bernard Katz, who lived in Australia for only a few years between 1938 and 1946 (though during the Second World War he served in the RAAF as a radar officer).

Bibliography

ABC TV, 'Oliphant remembered as a truly great Australian', *7:30 Report*, 17 July 2000, television program.
——, 'Brilliant creatures: Germaine, Clive, Barry & Bob', 23 September 2014, television program, <iview.abc.net.au/show/brilliant-creatures-germaine-clive-barry-and-bob>, accessed 25 November 2021.
Abraham, Edward, 'Howard Walter Florey, Baron Florey of Adelaide and Marston, 1898–1968', *Biographical Memoirs of Fellows of the Royal Society*, vol. 17, 1 November 1971, pp. 255–302.
——, 'Oxford, Howard Florey and World War II' in Carol L Moberg and Zanvil A Cohn (eds), *Launching the Antibiotic Era: Personal Accounts of the Discovery and Use of the First Antibiotics*, Rockefeller University Press, New York, 1990, pp. 19–30.
Abraham, EP, E Chain, CM Fletcher, HW Florey, AD Gardner, NG Heatley & MA Jennings, 'Further observations on penicillin', *Lancet*, vol. 238, 16 August 1941, pp. 177–89.
Ackerman, Evan, 'A brief history of the microwave oven: where the "radar" in Raytheon's Radarange came from', *IEEE Spectrum*, 30 September 2016, <spectrum.ieee.org/a-brief-history-of-the-microwave-oven>, accessed 25 November 2021.
Adelman, Jeremy, *Worldly Philosopher: The Odyssey of Albert O. Hirschman*, Princeton University Press, Princeton, 2013.
Allen, Thomas B & Norman Polmar, *Codename Downfall: The Secret Plan to Invade Japan*, Headline Book Publishing, London, 1995.
Alvarez, Luis W, *Adventures of a Physicist*, Basic Books, New York, 1987.
Aminov, Rustam I, 'A brief history of the antibiotic era: lessons learned and challenges for the future', *Frontiers in Microbiology*, vol. 1, December 2010, pp. 1–7.

Anderson, Dean, *Praise the Lord and Pass the Penicillin: Memoir of a Combat Medic in the Pacific in World War II*, McFarland, Jefferson, N.C., 2003.

Arnold, Lorna, 'The history of nuclear weapons: The Frisch–Peierls memorandum on the possible construction of atomic bombs of February 1940', *Cold War History*, vol. 3, no. 3, 2003, pp. 111–26.

Atkinson, Rick, *The Day Of Battle: The War in Sicily and Italy, 1943–1944*, Henry Holt, New York, 2007.

Atomic Heritage Foundation, in partnership with the National Museum of Nuclear Science & History, 'Debate over the Japanese surrender', <www.atomicheritage.org/history/debate-over-japanese-surrender>, accessed 13 January 2022.

——, 'Japanese atomic bomb project', <www.atomicheritage.org/history/japanese-atomic-bomb-project>, accessed 28 November 2021.

——, 'Manhattan Project history: Britain's early input – 1940–41', <www.atomicheritage.org/history/britains-early-input-1940-41>, accessed 25 November 2021.

Baime, AJ, *The Arsenal of Democracy: FDR, Detroit, and an Epic Quest to Arm an America at War*, Houghton Mifflin Harcourt, New York, 2014.

Baggott, Jim, *Atomic: The First War of Physics and the Secret History of the Atom Bomb, 1939–1949*, Icon, London, 2015.

Bahcall, Safi, 'The history of Pfizer and penicillin, and lessons for coronavirus', *Wall Street Journal*, 20 March 2020, <www.wsj.com/articles/the-history-of-pfizer-and-penicillin-and-lessons-for-coronavirus-11584723787>, accessed 14 December 2021.

Baldwin, Stanley, House of Commons Debates, 10 November 1932, <hansard.millbanksystems.com/commons/1932/nov/10/international-affairs#S5CV0270P0_19321110_HOC_284>, accessed 15 November 2021.

Banyard, MRC, 'Howard Walter Florey: A successful Australian', *Australian Science Mag*, issue 1, March 1987, pp. 8–12.

Barrett, Mike, 'Penicillin's first patient', *Mosaic*, 13 May 2018, <mosaicscience.com/story/penicillin-first-patient-history-albert-alexander-AMR-DRI/>, accessed 24 November 2021.

Baxter, James, *Scientists Against Time*, Little, Brown & Company, Boston, 1946.

BBC, 'Close down of television service for the duration of the war: 1 September 1939', History of the BBC, BBC website, <www.bbc.com/historyofthebbc/anniversaries/september/closedown-of-television>, accessed 24 November 2021.

——, 'The mould, the myth and the microbe', *Horizon*, 27 January 1986, television documentary.

Bibliography

——, 'On the right side of wrong', 17 February 2000, television documentary, BBC (Midlands).
Bennett, DCT, *Pathfinder: A War Autobiography*, Goodall Publications Limited, London, 1988.
Bernstein, Barton J, 'The uneasy alliance: Roosevelt, Churchill, and the atomic bomb, 1949–1945', *Western Political Quarterly*, vol. 29, no. 2, June 1976, pp. 202–30.
Bernstein, Jeremy, 'Profiles: I.I. Rabi', *New Yorker*, 20 October 1975, pp. 47–8.
——, 'A memorandum that changed the world', *American Journal of Physics*, vol. 79, no. 5, May 2011, pp. 440–6.
Beyerchen, Alan, 'On strategic goals as perceptual filters: Interwar responses to the potential of radar in Germany, the UK and the US', in O Blumtritt, H Petzold & W Aspray (eds), *Tracking the History of Radar*, IEEE-Rutgers Center for the History of Electrical Engineering and the Deutsches Museum Piscataway, New Jersey, 1994, pp. 267–283.
Bickel, Lennard, *Rise up to Life: A biography of Howard Walter Florey who made penicillin and gave it to the world*, Angus & Robertson, London, 1972.
——, *The Deadly Element: The Story of Uranium*, Stein & Day, New York, 1979.
Binnie, Anna, 'Oliphant, the father of atomic energy', *Journal & Proceedings of the Royal Society of New South Wales*, vol. 139, no. 419, 2006, pp. 11–22.
Blanchard, Yves, Gaspare Galati & Pietvan Genderen, 'The cavity magnetron: Not just a British invention', *IEEE Antennas and Propagation Magazine*, vol. 55, no. 5, October 2013, pp. 244–54.
Bleaney, Brebis, 'Sir Mark (Marcus Laurence Elwin) Oliphant, A.C., K.B.E.', *Biographical Memoirs of Fellows of the Royal Society*, vol. 47, November 2001, pp. 383–93.
Blythe, Max, 'Dr Paquita McMichael in interview with Dr Max Blythe, Edinburgh, 17 February 1998', Royal College of Physicians and Oxford Brookes University Medical Sciences Video Archive MSVA 178, <radar.brookes.ac.uk/radar/file/6b6a8fce-fa49–45f9–935a-6394f699f654/1/McMichael%2CP.pdf>, accessed 21 November 2021.
——, 'Professor Charles du Ve Florey in interview with Dr Max Blythe, Edinburgh, 17 February 1998', Royal College of Physicians and Oxford Brookes University Medical Sciences Video Archive MSVA 179, <radar.brookes.ac.uk/radar/file/7ea51825-d26c-4575–8973–985d58aec234/1/Florey%2CC.pdf>, accessed 13 January 2022.
Boot, Henry AH & John T Randall, 'Historical notes on the cavity magnetron', *IEEE Transactions on Electron Devices*, vol. 23, no. 7, 1976, pp. 724–9.

Bowden, ME, 'Old brew, new brew', *Distillations*, Science History Institute, 30 July 2018, <www.sciencehistory.org/distillations/old-brew-new-brew>, accessed 19 November 2021.

Bowen, EG, *Radar Days*, Hilger, Bristol, 1987.

Boyle, Andrew, *The Climate of Treason: Five Who Spied for Russia*, Hutchinson & Co Ltd, London, 1979.

Boyne, Walter J, 'The B-29's Battle of Kansas', *Air Force Magazine*, 1 February 2012, <www.airforcemag.com/article/0212b29/>, accessed 14 December 2021.

Brendon, Piers, *The Dark Valley: A Panorama of the 1930s*, Jonathan Cape, London, 2000.

Brittain, James E, 'The magnetron and the beginnings of the microwave age', *Physics Today*, vol. 38, no. 7, July 1985, pp. 60–7.

Bronowski, Jacob, *The Ascent of Man*, British Broadcasting Corporation, London, 1973.

Brown, Louis, 'Significant effects of radar on the Second World War', in O Blumtritt, H Petzold & W Aspray (eds), *Tracking the History of Radar*, IEEE-Rutgers Center for the History of Electrical Engineering and the Deutsches Museum Piscataway, New Jersey, 1994, pp. 121–35.

——, *Technical and Military Imperatives: A Radar History of World War II*, Institute of Physics Publishing, Bristol, 1999.

Bryson, Bill, *The Body: A Guide for Occupants*, Penguin, London, 2020.

Bud, Robert, *Penicillin: Triumph and Tragedy*, Oxford University Press, Oxford, 2007.

Buderi, Robert, *The Invention that Changed the World: The Story of Radar from War to Peace*, Little, Brown & Company, London, 1997.

Bundy, McGeorge, *Danger and Survival: Choices about the Bomb in the First Fifty Years*, Vintage Books, New York, 1990.

Bush, Vannevar, 'Foreword', in James Phinney Baxter, *Scientists Against Time*, 2nd edn, MIT Press, Cambridge, Mass., 1968.

Butcher, Harry C, *My Three Years with Eisenhower*, Simon & Schuster, New York, 1946.

Bynum, Bill, 'Shedding new light on the story of penicillin', *Lancet*, vol. 369, 16 June 2007, pp. 1991–2.

Campbell, John, 'Rutherford, transmutation and the proton', *CERN Courier*, 8 May 2019, <cerncourier.com/a/rutherford-transmutation-and-the-proton/>, accessed 20 January 2022.

Cannon, Michael, *The Roaring Days*, Today's Australia Publishing Company, Mornington, 1998.

Bibliography

Carpenter, Charles A, 'A "dramatic extravaganza" of the projected atomic age: *Wings over Europe*', *Modern Drama*, vol. 35, no. 4, 1992, pp. 552–61.

Carver, JH, RW Crompton, DG Ellyard, LU Hibbard & EK Inall, 'Marcus Laurence Elwin Oliphant 1901–2000', *Historical Records of Australian Science*, vol. 14, 2003, pp. 337–64.

Cathcart, Brian, *The Fly in the Cathedral: How a Group of Cambridge Scientists won the International Race to Split the Atom*, Farrar, Straus & Giroux, New York, 2004.

Chain, E, HW Florey, AD Gardner, NG Heatley, MA Jennings, J Orr-Ewing & AG Sanders, 'Penicillin as a chemotherapeutic agent', *Lancet*, vol. 236, no. 6104, 24 August 1940, pp. 226–8.

Chain, Ernst, 'Banquet speech: Ernst B. Chain's speech at the Nobel Banquet in Stockholm, December 10, 1945', <www.nobelprize.org/prizes/medicine/1945/chain/speech/>, accessed 24 November 2021.

——, 'Thirty years of penicillin therapy', *Proceedings of the Royal Society of London*, Vol. 179 (Series B), 1971, pp. 293–319.

Churchill, Winston, 'Fifty years hence', Macleans, 15 November 1931, <archive.macleans.ca/article/1931/11/15/fifty-years-hence>, accessed 14 November 2021.

——, House of Commons Debates, 13 May 1940, <api.parliament.uk/historic-hansard/commons/1940/may/13/his-majestys-government-1>, accessed 17 November 2021.

——, House of Commons Debates, 4 June 1940, <api.parliament.uk/historic-hansard/commons/1940/jun/04/war-situation>, accessed 17 November 2021.

——, House of Commons Debates, 18 June 1940, <api.parliament.uk/historic-hansard/commons/1940/jun/18/war-situation>, accessed 17 November 2021.

——, House of Commons Debates, 20 August 1940, <api.parliament.uk/historic-hansard/commons/1940/aug/20/war-situation>, accessed 17 November 2021.

——, *The Second World War*, vol. 1, *The Gathering Storm*, Houghton Mifflin, Boston, 1948.

——, *The Second World War*, vol. 2, *Their Finest Hour*, Houghton Mifflin, Boston, 1949.

Clark, Andrew, 'Young, free and confident', *Australian Financial Review*, 24 January 2015, p. 18.

Clark, Ronald W, *The Birth of the Bomb: The Untold Story of Britain's Part in the Weapon that Changed the World*, Phoenix House, London, 1961.

——, *The Rise of the Boffins*, Phoenix House, London, 1962.

——, *Tizard*, Methuen, London, 1965.

——, *The Life of Ernst Chain: Penicillin and Beyond*, Weidenfeld & Nicolson, London, 1985.

Clarke, CMH, *A History of Australia*, vol. 6, Melbourne University Press, Melbourne, 1987.
Cockburn, Stewart, 'Keith Hancock interviewed by Stewart Cockburn for the Mark Oliphant biography collection', 14 February 1980, National Library of Australia, ORAL TRC 889.
Cockburn, Stewart & David Ellyard, *Oliphant: The Life and Times of Sir Mark Oliphant*, Axiom Books, Adelaide, 1981.
Cohen, Richard, *Chasing the Sun: The Epic Story of the Star that Gives Us Life*, Simon & Schuster, London, 2011.
Compton, Arthur H, *Atomic Quest: A Personal Narrative*, Oxford University Press, London, 1956.
Conant, James, *My Several Lives: Memoirs of a Social Inventor*, Harper & Row, New York, 1970.
Conant, Jennet, *Tuxedo Park: A Wall Street Tycoon and the Secret Palace of Science that Changed the Course of World War II*, Simon & Schuster, New York, 2003.
Cornwall, John, *Hitler's Scientists: Science, War, and the Devil's Pact*, Penguin, New York, 2004.
Correll, John T, 'How the Luftwaffe lost the Battle of Britain', *Air Force Magazine*, 1 August 2008, <www.airforcemag.com/article/0808battle/>, accessed 18 November 2021.
Costello, John, *The Masks of Treachery*, William Morrow & Company, New York, 1988.
Cox, Brian, 'People of science with Brian Cox: Dame Sally Davies – Alexander Fleming and Howard Florey', <royalsociety.org/about-us/programmes/people-of-science/sally-davies-fleming-florey/>, accessed 25 November 2021.
Craven, Peter, 'Peter Conrad's *How the World was Won* explores the romance of America', *Weekend Australian*, 'Review' section, 22–23 November 2014, p. 22.
Dahl, Per F, *Heavy Water and the Wartime Race for Nuclear Energy*, Institute of Physics Publishing, Bristol, 1999.
Davis, Nuel Pharr, *Lawrence and Oppenheimer*, Fawcett, New York, 1968.
de Berg, Hazel, 'Howard Walter Florey interviewed by Hazel de Berg', 5 April 1967, *Hazel de Berg Collection*, National Library of Australia Oral History Transcript, <nla.gov.au/nla.obj-220872612>, accessed 13 January 2022.
——, 'Interview notes, Interview with Howard Florey, Tape 220–21', 5 April 1967.
Diesendorf, Mark, 'Is nuclear energy a possible solution to global warming?' Institute of Environmental Studies, UNSW, <www.ceem.unsw.edu.au/sites/default/files/uploads/publications/NukesSocialAlternativesMD.pdf>, accessed 13 January 2022.

Bibliography

Dimbleby, Jonathan, *The Battle of the Atlantic: How the Allies Won the War*, Penguin, London, 2016.
Doenitz, Karl, *Memoirs: Ten Years and Twenty Days*, Weidenfeld & Nicolson, London, 1959.
Doherty, Peter, 'Time to champion our tall poppies', *AQ: Australian Quarterly*, vol. 70, no. 6, Nov–Dec 1998, pp. 54–6.
Dowling, Henry F, *Fighting Infection: Conquests of the Twentieth Century*, Harvard University Press, Cambridge Mass., 1977.
Duncan, WGK & Roger Leonard, *The University of Adelaide, 1874–1974*, Rigby, Adelaide, 1973.
Edwards, Sam, 'Rudolph E Peierls', *Physics Today*, vol. 49, no. 2, 1996, pp. 74–5.
Ellyard, David, 'Mark Oliphant FRS and the Birmingham proton synchrotron', PhD thesis, University of New South Wales, December 2011, <unsworks.unsw.edu.au/fapi/datastream/unsworks:10945/SOURCE01?view=true>, accessed 25 November 2021.
Fairley, Peter, 'Why don't we have more nuclear power?', *MIT Technology Review*, 28 May 2015, <www.technologyreview.com/2015/05/28/167951/why-dont-we-have-more-nuclear-power/>, accessed 25 November 2021.
Farmelo, Graham, *Churchill's Bomb: How the United States Overtook Britain in the First Nuclear Arms Race*, Basic Books, New York, 2013.
'Fateful discovery almost forgotten', *Nature*, vol. 337, 9 February 1989, pp. 499–502.
Faulkner, William, Speech at the Nobel Banquet, 10 December 1950, <www.nobelprize.org/prizes/literature/1949/faulkner/speech/>, accessed 13 January 2022.
Fenner, Frank, 'Obituary: Howard Walter Florey: Baron of Adelaide and Marston', *Australian Journal of Science*, vol. 31, no. 1, 1968, pp. 37–9.
——, 'Howard Florey', *Microbiology Australia*, vol. 19, no. 3, July 1998, pp. 26–8.
Ferguson, Niall, *The War of the World: History's Age of Hatred*, Penguin Books, London, 2007.
Fijałek, Krzysztof, '1 Września 1939 r., Wieluń. Masakra na bezbronnym mieście', *Interia*, 1 September 2019, <wydarzenia.interia.pl/raporty/raport-zbrodnia-bez-kary/historie/news-1-wrzesnia-1939-r-wielun-masakra-na-bezbronnym-miescie,nId,3179824>, accessed 24 November 2021.
Fine, Norman, *Blind Bombing: How Microwave Radar Brought the Allies to D-Day and Victory in World War II*, University of Nebraska Press, Lincoln, 2019.
Fisher, David E, *A Race on the Edge of Time: Radar – The Decisive Weapon of World War II*, McGraw Hill, New York, 1989.
FitzSimons, Peter, *Nancy Wake: A Biography of Our Greatest War Heroine*, Harper Collins Publishers, Sydney, 2001.

Fleming, Alexander, 'On the antibacterial action of cultures of a penicillium, with special reference to their use in the isolation of *B. Influenzae*', *British Journal of Experimental Pathology*, vol. 10, no. 3, June 1929, pp. 226–36.

Florey, H, 'Penicillin', *Nobel Lecture*, 11 December 1945, <www.nobelprize.org/prizes/medicine/1945/florey/lecture/>, accessed 15 November 2021.

Florey, HW, E Chain, NG Heatley, MA Jennings, AG Saunders, EP Abraham & ME Florey, *Antibiotics*, Oxford University Press, London, 1949.

Foote, Shelby, 'The art of fiction no. 158', *Paris Review*, issue 151, Summer 1999, <www.theparisreview.org/interviews/931/the-art-of-fiction-no-158-shelby-foote>, accessed 13 January 2022.

Foster, SC & Margaret M Varghese, *The Making of the Australian National University, 1946–1996*, Allen & Unwin, Sydney, 1996.

Fox, Robin Lane, *The Invention of Medicine: From Homer to Hippocrates*, Penguin, London, 2020.

Frisch, Otto, 'How it all began', *Physics Today*, vol. 20, no. 11, November 1967, pp. 43–8.

——, *What Little I Remember*, Cambridge University Press, New York, 1979.

Funk, Duane J, Joseph E Parillo & Anand Kumar, 'Sepsis and septic shock', *Critical Care Clinics*, vol. 25, February 2009, pp. 83–101.

Fussell, Paul, 'Thank God for the atom bomb', in *Thank God for the Atom Bomb and Other Essays*, Summit Books, New York, 1988, pp. 13–37.

Gaddis, John Lewis, *The Long Peace: Inquiries into the History of the Cold War*, Oxford University Press, New York, 1987.

Galati, Gaspare & Piet van Gendaren, 'Introduction to the special section on some less-well-known contributions to the development of radar: From its early conception until just after the Second World War', *Radio Science Bulletin*, no. 358, September 2016, pp. 12–6.

Gardiner, Juliet, *Wartime: Britain 1939–1945*, Headline, London, 2005.

Gaudin, Sharon, 'Titanic was high-tech marvel of its time', *Computerworld*, 14 April 2012, <www.computerworld.com/article/2503092/titanic-was-high-tech-marvel-of-its-time.html?page=2>, accessed 25 November 2021.

Gaynes, Robert, 'The discovery of penicillin – New insights after more than 75 years of clinical use', *Emerging Infectious Diseases*, vol. 23, no. 5, May 2017, pp. 849–53.

Gill, Trevor, 'Sir Mark: I would prefer to die', *Sunday Mail*, 21 July 1985, p. 47.

Goldberg, Stanley, 'Inventing a climate of opinion: Vannevar Bush and the decision to build the bomb', *ISIS*, vol. 83, no. 3, September 1992, pp. 429–52.

Goldsworthy, Peter D & Alexander C McFarlane, 'Howard Florey, Alexander Fleming and the fairy tale of penicillin', *Medical Journal of Australia*, vol. 176, 18 February 2002, pp. 178–80.

Bibliography

Goodwin, Doris Kearns, *No Ordinary Time: Franklin and Eleanor Roosevelt: The Home Front in World War II*, Simon & Schuster, New York, 1995.

Gould, Kate, 'Antibiotics: From prehistory to the present day', *Journal of Antimicrobal Chemotherapy*, vol. 71, 2016, pp. 572–5.

Gowing, Margaret, *Britain and Atomic Energy, 1939–1945*, Macmillan, London, 1964.

Goyette, Dolores, 'The surgical legacy of World War II, Part 2: The age of antibiotics', *The Surgical Technologist*, June 2017, pp. 257–64.

Gross, Daniel P & Bhaven N Sampat, *Inventing the Endless Frontier: The Effects of the World War II Research Effort on Post-war Innovation*, Working Paper 27375, National Bureau of Economic Research, June 2020, <www.nber.org/papers/w27375>, accessed 27 January 2022.

Groueff, Stephane, 'J Robert Oppenheimer interviewed by Stephane Groueff', 1965, <www.manhattanprojectvoices.org/oral-histories/j-robert-oppenheimers-interview>, accessed 26 November 2021.

Groves, Leslie R, *Now It Can Be Told: The Story of the Manhattan Project*, Andre Deutsch, London, 1963.

Guerlac, Henry, *Radar in World War II*, American Institute of Physics, New York, 1987.

Guerlac, Henry & Marie Boas, 'The radar war against the U-boat', *Military Affairs*, vol. 14, no. 2, Summer 1950, pp. 99–111.

Ham, Paul, *Hiroshima and Nagasaki*, Harper Collins, Sydney, 2011.

Hancock, WK, *Country and Calling*, Faber & Faber, London, 1954.

Harford, Tim, 'How the search for a "death ray" led to radar', *50 Things that Made the Modern Economy*, 9 October 2017, radio broadcast, BBC world service, <www.bbc.com/news/business-41188464>, accessed 15 November 2021.

Hargadon, Andrew, 'It's not about the idea', *TEDxSacramento*, 12 December 2012, <www.youtube.com/watch?v=wtsdkq97AMw>, accessed 8 December 2021.

Harris, Henry, 'Howard Florey and the development of penicillin', *Notes and Records: Royal Society Journal of the History of Science*, vol. 53, no. 2, 1999, pp. 243–52.

Hasluck, Alexandra (ed.), *Audrey Tennyson's Vice Regal Days: The Australian Letters of Audrey Lady Tennyson to her mother Zacintha Boyle, 1899–1903*, National Library of Australia, Canberra, 1978.

Haygarth, Stephanie, 'A maker of miracles', *AQ: Australian Quarterly*, September–October 1997, pp. 32–8.

Heagney, Brenda, *Half a Century of Penicillin: An Australian Perspective*, Royal Australasian College of Physicians, Sydney, 1991.

Heatley, Norman, 'Penicillin and luck', in Carol L Moberg & Zanvil A Cohn (eds), *Launching the Antibiotic Era: Personal Accounts of the Discovery and Use of the First Antibiotics*, Rockefeller University Press, New York, 1990, pp. 31–41.

Herman, Arthur, *Freedom's Forge: How American Business Produced Victory in World War II*, Random House, New York, 2013.

Hewlett, Richard G & Oscar E Anderson, Jr. *History of the United States Atomic Energy Commission*, vol. 1, *The New World, 1939–1946*, Pennsylvania State University Press, University Park, 1962.

Hind, Angela, 'Briefcase that "changed the world"', *The World in a Briefcase*, 5 February 2007, radio program, Pier Productions, BBC Radio 4, <news.bbc.co.uk/2/hi/science/nature/6331897.stm>, accessed 18 November 2021.

Hiltzik, Michael, 'The man who saved the Manhattan Project', History News Network, 20 September 2015, <historynewsnetwork.org/article/160528>, accessed 20 January 2022.

———, *Big Science: Ernest Lawrence and the Invention that Launched the Military-Industrial Complex*, Simon & Schuster, New York, 2015.

Hobby, Gladys L, *Penicillin: Meeting the Challenge*, Yale University Press, New Haven, 1985.

Holden, Darren, 'The indiscretion of Mark Oliphant: How an Australian kick-started the American atomic bomb project', *Historical Records of Australian Science*, vol. 29, no. 1, 2018, pp. 28–35.

———, '"On the Oliphant Design, Now Sound the Blast"; How Mark Oliphant secretly warned of America's post-war intentions of an atomic monopoly', *Historical Records of Australian Science*, vol. 29, no. 2, 2018, pp. 130–7.

———, 'Mark Oliphant and the Invisible College of the Peaceful Atom', PhD thesis, University of Notre Dame Australia, 2019, <researchonline.nd.edu.au/cgi/viewcontent.cgi?article=1278&context=theses>, accessed 26 November 2021.

Holman, Brett, 'The air panic of 1935: British press opinion between disarmament and rearmament', *Journal of Contemporary History*, vol. 46, no. 2, 2011, pp. 288–307.

———, *The Next War in the Air: Britain's Fear of the Bomber, 1908–1941*, Routledge, London, 2014.

Home, RW, 'The problem of intellectual isolation in scientific life: WHT Bragg and the Australian scientific community, 1886–1909', *Historical Records of Australian Science*, vol. 6, no. 1, 1984, pp. 19–30.

Hopkins, Eric, *Birmingham: The First Manufacturing Town in the World, 1760–1840*, Weidenfeld & Nicolson, London, 1989.

Hughes, Robin, *Australian Lives: Stories of Twentieth Century Australians*, Angus & Robertson, Sydney, 1996.

Bibliography

Hutchings, Matthew, Andrew W Truman & Barrie Wilkinson, 'Antibiotics: Past, present and future', *Current Opinion in Microbiology*, vol. 51, October 2019, pp. 72–80.

Isherwood, Christopher, *Lions and Shadows: An Education in the Twenties*, Methuen, London, 1985.

Jenkins, Roy, *Churchill*, Macmillan, London, 2001.

Jonas, Gerald, *The Circuit Riders: Rockefeller Money and the Rise of Modern Science*, WW Norton, New York, 1989.

Jones, RV, 'Sir Robert Alexander Watson-Watt', *Oxford Dictionary of National Bibliography*, <www.oxforddnb.com/view/10.1093/ref:odnb/9780198614128.001.0001/odnb-9780198614128-e-9000022?rskey=NMgm8z&result=1>, accessed 12 January 2022.

——, *Most Secret War: British Scientific Intelligence 1939–1945*, Coronet, London, 1979.

Jones, Vincent C, *Manhattan: The Army and the Atomic Bomb*, Center of Military History United States Army, Washington, 1985.

Judkins, Phil, 'Making visions into power: Power struggles and personality clashes in British radar, 1935–1941', *International Journal for the History of Engineering and Technology*, vol. 82, no. 1, 2012, pp. 93–124.

Jungk, Robert, *Brighter than a Thousand Suns*, Harcourt, New York, 1958.

Kahn, EJ, 'The universal drink: The making-and selling-of Coca-Cola', *New Yorker*, 14 February 1959, <www.newyorker.com/magazine/1959/02/14/the-universal-drink>, accessed 30 November 2021.

Kaiser, Walter, 'A case study in the relationship of history of technology and of general history: British radar technology and Neville Chamberlain's appeasement', *Icon*, vol. 2, 1996, pp. 29–52.

Keeble, Kathryn M, 'Frankenstein's machine: Redressing Mark Oliphant's scientific reputation', *Historical Records of Australian Science*, vol. 29, no. 2, 2018, pp. 122–8.

——, 'General Groves's "inevitable Wwar with Russia": Joseph Rotblat's and Mark Oliphant's existential crises', *Journal of Australian Studies*, 15 October 2021, <doi.org/10.1080/14443058.2021.1987293>, accessed 20 January 2022.

Kellerhof, Sven Felix, 'Hitler's physicists and the bomb', 1 February 2015, <www.ozy.com/true-and-stories/hitlers-physicists-and-the-bomb/37426/>, accessed 28 November 2021.

Kelly, Cindy, 'Lawrence Bartell interviewed by Cindy Kelly, Ann Arbor, Michigan', 9 May 2013, <www.manhattanprojectvoices.org/oral-histories/lawrence-bartells-interview>, accessed 29 November, 2021.

Kelly, Cynthia C, *The Manhattan Project: The Birth of the Atomic Bomb in the Words of Its Creators, Eyewitnesses, and Historians*, Black Dog & Leventhal Publishers, New York, 2020.

Kendal, Brian, 'An overview of the development and introduction of ground radar to 1945', *Journal of Navigation*, vol. 56, no. 3, September 2003, pp. 343–52.

Kennedy, Paul, *Engineers of Victory: The Problem Solvers Who Turned the Tide in the Second World War*, Penguin, London, 2014.

Kevles, Daniel, *The Physicists*, Alfred A Knopf, New York, 1978.

Khatami, Ameneh, Philip N Britton, Glendon Farrow, Megan Phelps & Alyson Kakakios, 'Meningitis and the military: The remarkable story of the first use of penicillin in Australia', *MJA*, vol. 213, no. 11, 14 December 2020, pp. 508–10.

Klein, Maury, *A Call to Arms: Mobilising America for World War II*, Bloomsbury Press, New York, 2013.

Kotulak, Ronald, 'Sorry Mary, you're no hero after all: Real penicillin star is unknown housewife', *The Dispatch* (Moline, Illinois), 10 October 1976, <clickamericana.com/topics/health-medicine/how-penicillin-was-discovered>, accessed 8 December 2021.

Landas, Marc, *Cold War Resistance: The International Struggle Over Antibiotics*, Potomac Books, Lincoln, 2020.

Larkin, John, 'Genius and the genie: Sir Mark Oliphant on the atomic bomb, Interview by John Larkin', *Age*, 8 August 1985, p. 11.

Larson, Clarence, *Pioneers of Science and Technology: Interview with Mark Oliphant*, Washington DC, 11 March 1988, <www.youtube.com/watch?v=z0QaXvLsR9A>, accessed 10 November 2021.

'Last Year's Graduates', *Adelaide Medical Students Society Review*, vol. XIII, no. 21, July 1922, p. 50.

Latham, Colin & Anne Stobbs, *Pioneers of Radar*, Sutton Publishing, Thrupp, Stroud, 1999.

Laucht, Christoph, *Elemental Germans: Klaus Fuchs, Rudolf Peierls and the Making of British Nuclear Culture, 1939–59*, Palgrave Macmillan, London, 2012.

Lax, Eric, *The Mold in Dr. Florey's Coat: The Story of the Penicillin Miracle*, Henry Holt & Company, New York, 2004.

Lee, Sabine, 'Birmingham – London – Los Alamos – Hiroshima: Britain and the atomic bomb', *Midland History*, vol. 21, no. 1, 2002, pp. 146–64.

——, '"In no sense vital and actually not even important"? Reality and perception of Britain's contribution to the development of nuclear weapons', *Contemporary British History*, vol. 20, no. 2, 2006, pp. 159–85.

Bibliography

——, '"Crucial? Helpful? Practically nil?" Reality and perception of Britain's contribution to the development of nuclear weapons during the Second World War', *Diplomacy and Statecraft*, vol. 33, no. 1, 6 May 2022, pp. 19–40.

Levine, Joshua, *Dunkirk: The History Behind the Major Motion Picture*, William Collins, London, 2017.

Liebenau, Jonathan, 'The British success with penicillin', *Social Studies of Science*, vol. 17, no. 1, 1987, pp. 69–86.

Liljestrand, G, 'Award ceremony speech: Presentation speech by Professor G Liljestrand, member of the Staff of Professors of the Royal Caroline Institute, on December 10, 1945', < www.nobelprize.org/prizes/medicine/1945/ceremony-speech/>, accessed 24 November 2021.

Lochery, Neill, *Lisbon: War in the Shadows of the City of Light, 1939–1945*, Scribe, Melbourne, 2011.

Loftis, Larry, *Into the Lion's Mouth: The True Story of Dusko Popov: World War II Spy, Patriot and the Real-life Inspiration for James Bond*, Berkley Caliber, New York, 2016.

Logevall, Fredrik, *JFK*, vol. I, *1917–1956*, Viking, London, 2020.

Longmate, Norman, *How We Lived Then: A History of Everyday Life During the Second World War*, Pimlico, London, 2002.

Loudon, Irvine, 'Deaths in childbed from the eighteenth century to 1935', *Medical History*, vol. 30, no. 1, 1986, pp. 1–41.

'Luftwaffe air war Poland 1939', *Weapons and Warfare*, 4 May 2019, <weaponsandwarfare.com/2019/05/04/luftwaffe-air-war-poland-1939/>, accessed 24 November 2021.

Lukacs, John, *Five Days in London: May 1940*, Scribe, Sydney, 2001.

MacDonogh, Giles, *After the Reich: From the Liberation of Vienna to the Berlin Airlift*, John Murray, London, 2008.

Macfarlane, Gwyn, *Howard Florey: The Making of a Great Scientist*, Oxford University Press, Oxford, 1979.

——, *Alexander Fleming: The Man and the Myth*, Chatto & Windus, London, 1984.

McNeill, William H, *Plagues and People*, Penguin, London, 1979.

Manchester, William, *The Last Lion*, vol. 2, *Winston Spencer Churchill – Alone, 1932–1940*, Dell, New York, 1989.

——, & Paul Reid, *The Last Lion*, vol. 3, *Winston Spencer Churchill – Defender of the Realm, 1940–1965*, Little, Brown & Company, New York, 2012.

Marsh, Allison, 'From World War II Radar to Microwave popcorn, the cavity magnetron was there', *IEEE Spectrum*, 31 October 2018, <spectrum.ieee.org/magnetron>, accessed 2 December 2021.

Medwar, Jean & David Pyke, *Hitler's Gift: Scientists Who Fled Nazi Germany*, Judy Piatkus, London, 2000.
Moore, Patrick, *Confessions of a Greenpeace Dropout: The Making of a Sensible Environmentalist*, Beatty Street Publishing, Vancouver, 2010.
Morris, James, *Pax Britannica: The Climax of Empire*, Faber & Faber, London, 1968.
Neushul, Peter, 'Science, government, and the mass production of penicillin, *Journal of the History of Medicine and Allied Sciences*, vol. 48, October 1993, pp. 371–95.
Nicolson, Harold, *Diaries and Letters: 1939–1945*, Collins, London, 1967.
O'Connell, Robert, *Soul of the Sword: An Illustrated History of Weaponry and Warfare from Prehistory to the Present*, Free Press, New York, 2002.
Ohler, Norman, *Blitzed: Drugs in Nazi Germany*, Penguin, London, 2017.
Oliphant, Mark, 'The Cambridge year: The seventh Cecil Eddy Memorial oration, 1964', *The Radiographer*, vol. II, no. 2, 1964, pp. 6–9.
——, 'The two Ernests – Part 1', *Physics Today*, vol. 19, no. 9, 1966, pp. 33–49.
——, 'Some recollections of Lord Florey', *ANU News*, April 1969, pp. 8–9.
——, *Rutherford – Recollections of the Cambridge Days*, Elsevier, London, 1972.
——, 'Looking back: Sir Mark Oliphant in conversation with David Ellyard', in Macfarlane Burnet & Mark Oliphant, *Sir Frank Macfarlane Burnet on Ageing & Looking Back by Sir Mark Oliphant*, Australian Broadcasting Commission, Sydney, 1979.
——, 'The beginning: Chadwick and the neutron', *Bulletin of the Atomic Scientists*, vol. 38, no. 10, 1982, pp. 14–18.
—— & Lord Penney, 'John Douglas Cockcroft, 1897–1967', *Biographical Memoirs of Fellows of the Royal Society*, vol. 14, November 1968, pp. 139–88.
Parsons, Coleman O, 'Mark Twain in Adelaide, South Australia', *Mark Twain Journal*, vol. 21, no. 3, 1983, pp. 51–4.
Peierls, Rudolf, *Bird of Passage: Recollections of a Physicist*, Princeton University Press, Princeton, 1985.
'Penicillin – Like lottery luck', *Courier Mail*, undated, Royal Society, M1944, Series 371, AJCP ref: 74_nla.obj-739916847-m.
Perry, Roland, 'The professor and the atom bomb', *Age*, 22 July 2000, 'News Extra', p. 2.
Persson, Sheryl, *Smallpox, Syphilis and Salvation: Medical Breakthroughs that Changed the World*, Exisle Publishing, Sydney, 2009.
Phelps, Stephen, *The Tizard Mission: The Top-Secret Operation that Changed the Course of World War II*, Westholme, Yardley, 2010.
Phillips, AA, 'The cultural cringe', *Meanjin*, vol. 9, no. 4, 1950, pp. 299–302.

Bibliography

Pielke, Roger, 'In retrospect: *Science – The Endless Frontier*', *Nature*, vol. 466, 19 August 2010, pp. 922–3.

Prange, Gordon C, *At Dawn We Slept: The Untold Story of Pearl Harbor*, Penguin, New York, 1991.

Preston, Diana, *Before the Fall-Out: The Human Chain Reaction from Marie Curie to Hiroshima*, Corgi Books, London, 2006.

Rabbitt Roff, Sue, 'Was Sir Mark Oliphant Australia's—and Britain's—J. Robert Oppenheimer?', *Meanjin Quarterly*, 22 January 2019, <meanjin.com.au/uncategorised/was-sir-mark-oliphant-australias-and-britains-j-robert-oppenheimer/>, accessed 14 December 2021.

——, 'Making the jitterbug work: Marcus Oliphant and the Manhattan Project', 30 May 2019, <www.atomicheritage.org/history/making-jitterbug-work-marcus-oliphant-and-manhattan-project>, accessed 6 January 2022.

Ramsey, Andrew, *The Basis of Everything: Rutherford, Oliphant and the Coming of the Atomic Bomb*, Harper Collins, Sydney, 2019.

Randall, JT, 'The cavity magnetron', *Proceedings of the Physical Society*, vol. 58, 1946, pp. 247–52.

Reader, Joseph & Charles Clark, '1932, a watershed year in nuclear physics', *Physics Today*, vol. 66, no. 3, March 2013, pp. 44–9.

Reynolds, David, *The Long Shadow: The Great War and the Twentieth Century*, Simon & Schuster, London, 2013.

Rhodes, Richard, *The Making of the Atomic Bomb*, Simon & Schuster, New York, 1986.

——, *Energy: A Human History*, Simon & Schuster, New York, 2018.

Richards, AN, 'Production of penicillin in the United States (1941–1946)', *Nature*, no. 4918, February 1, 1964, pp. 441–5.

Richardson, Ruth, 'Heatley's vessel', *Lancet*, vol. 357, no. 9264, 21 April 2001, p. 1298.

Rigden, John S, *Rabi: Scientist and Citizen*, Basic Books, New York, 1987.

Roberts, Conrad, 'The white Oliphant?', *Bulletin*, 25 January 1961, pp. 12–15.

Root-Bernstein, Robert, 'Howard Florey: Photographer, cinematographer and Sunday painter', *Leonardo*, vol. 42, no. 3, 2009, p. 265.

Rose, Norman, *Churchill: An Unruly Life*, Tauris Parke, New York, 2009.

Rosen, Raphael Chayim, 'Under the radar: Physics, engineering, and the distortion of a World War Two legacy', honors thesis, Harvard University, Cambridge, Mass., March 2006, <canvas.harvard.edu/files/4329530/download?download_frd=1>, accessed 18 November 2021.

Rosen, William, *Miracle Cure: The Creation of Antibiotics and the Birth of Modern Medicine*, Penguin Books, New York, 2017.

Roser, Max, 'Mortality in the past – around half died as children', *Our World in Data*, 11 June 2019, <ourworldindata.org/child-mortality-in-the-past>, accessed 26 January 2022.

Rowe, AP, *One Story of Radar*, Cambridge University Press, Cambridge, 1948.

Rubin, Ronald P, 'A brief history of great discoveries in pharmacology: In celebration of the centennial anniversary of the founding of the American Society of Pharmacology and Experimental Therapeutics', *Pharmacological Review*, vol. 59, no. 4, 2007, pp. 289–359.

Saad, TA, 'The story of the MIT radiation laboratory', *IEEE Aerospace and Electronic Systems Magazine*, vol. 5, no. 10, 1990, pp. 46–51.

Shellenberger, Michael, *Apocalypse Never: Why Environmental Alarmism Harms Us All*, Harper Collins, New York, 2020.

Shannon, John G & Paul M Moser, 'A history of US Navy periscope detection radar: Sensor design and development', Office of Naval Research, 2014, <apps.dtic.mil/dtic/tr/fulltext/u2/1003753.pdf accessed 1 December 2021>, accessed 26 November 2021.

Sharma, Gilbert & Jonathan Reinarz, 'Allied intelligence reports on wartime German penicillin research and production', *Historical Studies in the Physical and Biological Sciences*, vol. 32, no. 2, 2002, pp. 347–61.

Shifman, M, 'The beginning of the nuclear age', pp. 1–35, <www.semanticscholar.org/paper/The-Beginning-of-the-Nuclear-Age-Shifman/e10709fc56d91ac84b5ff50b1ec5ed320d46de07>, accessed 25 January 2022.

Smyth, Henry De Wolf, *Atomic Energy for Military Purposes*, Maple Press, York, Pa., 1945.

Snow, CP, *The Physicists*, Macmillan, London, 1981.

Stewart, Irvin, *Organizing Scientific Research for War: The Administrative History of the Office of Scientific Research and Development*, Atlantic-Little Brown, Boston, 1948.

Swords, SS, *Technical History of the Beginnings of Radar*, Institution of Engineering and Technology, London, 2008.

Szasz, Ferenc Morton, *British Scientists and the Manhattan Project: The Los Alamos Years*, St Martin's Press, New York, 1992.

Talty, Stephan, *The Illustrious Dead: The Terrifying Story of How Typhus Killed Napoleon's Greatest Army*, Crown Publishers, New York, 2009.

Taylor, AJP, *The Second World War: An Illustrated History* [1975] in *The Second World War and Its Aftermath*, Folio Society, London, 1998.

Taylor, Harold, 'What kind of person in James Bryant Conant? What has been his influence?', *New York Times*, 22 March 1970, p. 288, <www.nytimes.com/1970/03/22/archives/what-kind-of-person-is-james-bryant-conant-what-has-been-his.html>, accessed 12 January 2022.

Bibliography

Taylor, John, 'How World War I changed British universities forever',
The Conversation, 9 November 2018, <theconversation.com/how-world-war-i-changed-british-universities-forever-106104>, accessed 11 November 2021.

Tibbets, Paul W, *Return of the Enola Gay*, Enola Gay Remembered, New Hope, Pa., 1998.

Trim, Richard M, 'A brief history of the development of radar in Great Britain up to 1945', *Measurement + Control*, vol. 35, December 2005, pp. 299–301.

United States War Department, *Biennial Reports of the Chief of the United States Army to the Secretary of War: 1 July 1939 – 30 June 1945*, Center of Military History, United States Army, Washington D.C., 1996, <history.army.mil/html/books/070/70-57/CMH_Pub_70-57.pdf>, accessed 29 November 2021.

University of Adelaide, Faculty of Health and Medical Sciences, website, 'One Discovery That Changed The World', Florey 120 Anniversary, 2018, <health.adelaide.edu.au/florey120anniversary/one-discovery-that-changed-the-world>, accessed 25 November 2021.

VanDeMark, Brian, *Pandora's Keepers: Nine Men and the Atomic Bomb*, Back Bay Books, New York, 2005.

Wainwright, Milton, 'Moulds in folk medicine', *Folklore*, vol. 100, no. 2, 1989, pp. 162–6.

——, 'Hitler's penicillin', *Perspectives in Biology and Medicine*, vol. 47, no. 2, 2004, pp. 189–98.

Watson-Watt, Robert, *Three Steps to Victory: A Personal Account By Radar's Greatest Pioneer*, Odhams Press, London, 1957.

Waugh, Evelyn, *Brideshead Revisited*, Penguin Classics, London, 2000.

Weart, Spencer & Gertrude Szilard (eds), *Leo Szilard: His Version of the Facts (Selected Recollections and Correspondence)*, MIT Press, Cambridge, Mass., 1978.

Weiner, Charles, 'Oral histories: Otto Frisch', Niels Bohr Library and Archives, American Institute of Physics, 3 May 1967, <www.aip.org/history-programs/niels-bohr-library/oral-histories/4616>, accessed 15 November 2021.

——, 'Oral histories: Mark Oliphant', Niels Bohr Library & Archives, American Institute of Physics, 3 November 1971, <www.aip.org/history-programs/niels-bohr-library/oral-histories/4805>, accessed 21 November 2021.

Wells, HG, *The War of the Worlds*, William Heinemann, London, 1998.

Wernert, Gregory T, 'Howard Florey: The man behind penicillin', *Chemistry in Australia*, vol. 65, no. 8, September 1998, pp. 21–5.

Wheeler, John, 'Mechanism of fission', *Physics Today*, vol. 20, no. 11, November 1967, pp. 49–52.

White, Michael, *The Fruits of War: How Military Conflict Accelerates Technology*, Simon & Schuster, London, 2005.
White, Patrick, 'Prodigal son', in Imre Salusinszky (ed.), *The Oxford Book of Australian Essays*, Oxford University Press, Melbourne, 1997, pp. 125–8.
Williams, Robyn, 'Sir Mark Oliphant', *Life and Times*, 1985, radio program, ABC Radio National, <www.abc.net.au/rn/legacy/programs/lifeandtimes/stories/2009/2588019.htm>, accessed 11 November 2021.
Williams, Trevor I, *Howard Florey: Penicillin and After*, Oxford University Press, Oxford, 1984.
Wilson, AN, *After the Victorians: 1901–1953*, Hutchinson, London, 2005.
Wilson, EJN, 'Fifty years of synchrotron', 5th European Particle Accelerator Conference, Sitges, Barcelona, 10–14 June 1996, <accelconf.web.cern.ch/e96/papers/orals/frx04a.pdf>, accessed 25 November 2021.
Wilson, Ward, 'The myth of nuclear deterrence', *Nonproliferation Review*, vol. 15, no. 3, November 2008, pp. 421–39.
World Health Organization, 'Tuberculosis: Key facts', *World Health Organization*, 14 October 2021, <www.who.int/news-room/fact-sheets/detail/tuberculosis>, accessed 25 November 2021.
York, Barry, 'Howard Florey and the development of penicillin', *National Library of Australia News*, vol. XI, no. 12, September 2001, pp. 18–20.
Zimmerman, David, 'The Tizard Mission and the development of the atomic bomb', *War in History*, vol. 2, no. 3, 1995, pp. 259–273.
——, *Top Secret Exchange: The Tizard Mission and the Scientific War*, Alan Sutton Publishing Limited, Phoenix Mill, 1996.

Picture credits

Page i: Author photograph by Kris Anderson; **page 1:** Courtesy of University of Adelaide Library, Rare Books and Manuscripts. Donor: Dr. Joan Gardner; **page 2:** Monica Oliphant; **page 3 top:** Science Museum London / Science and Society Picture Library, CC BY-SA 2.0, via Wikimedia Commons; **page 3 bottom:** photograph by Nordisk Pressefoto, Niels Bohr Institute, courtesy of AIP Emilio Segrè Visual Archives, Fermi Film Collection; **page 4:** Courtesy of University of Adelaide Library, Rare Books and Manuscripts. Donor: Dr. Joan Gardner; **page 5:** Lawrence Berkeley National Laboratory; **page 6 top:** Ministry of Information official photographer, via Wikimedia Commons; **page 6 bottom:** Alamy; **page 7 top:** Alamy; **page 7 bottom:** Photographer unknown; **page 8 top:** Blackstone Studios, CC BY 4.0, via Wikimedia Commons; **page 8 bottom:** Courtesy of University of Adelaide Library, Rare Books and Manuscripts; **page 9 top:** Alamy; **page 9 bottom:** Lawrence Berkeley National Laboratory; **page 10 top:** Dunn School of Pathology, Oxford University; **page 10 bottom:** USDA Agricultural Research Service/National Center for Agricultural Utilization Research; **page 11 top:** Dunn School of Pathology, Oxford University; **page 11 bottom:** Alamy; **page 12 top:** Royal Air Force official photographer, via Wikimedia Commons; **page 12 bottom:** The Official CTBTO Photostream, CC BY 2.0, via Wikimedia Commons; **page 13 top:** Alamy; **page 13 bottom:** Alamy; **page 14:** photographer unknown; **page 15:** Australian National University Archives; **page 16 top:** Australian National University Archives; **page 16 bottom:** Charles Florey

Index

A
Abraham, Edward 123, 275
Adelaide 1, 11–13, 16–23
Adrian, Edgar 33
Akers, Elva 192–93
Alexander, Albert 193, 203
Alexander, General Sir Harold 280
Allier, Jacques 142
Alvarez, Luis 179
Anderson, Sir John 188, 295
antibacterial substances 38–39, 65, 67, 84, 86–87, 151, 271–72, 320
  see also penicillin
antibiotics 63–64, 319–20, 337
  see also penicillin
Appleton, Sir Edward 243, 292–93
Army Signals Corp (US) 253
Arnold, Lorna 134
Atlantic Charter 225
atomic bomb 237–38
  Advisory Committee on Uranium (US) 180–81, 187, 217–20, 222–24, 230, 239, 240, 244–45, 344
  cost of Manhattan Project 295–96, 309
  cyclotrons 91–92, 96, 108, 167, 175, 187, 229, 230, 240, 298, 300, 305, 323, 328
  electromagnetic isotope separation 298–300
  espionage 327
  fission research 77, 125–33, 229–30, 301
  Frisch/Peierls memorandum 131–36, 142, 171, 185, 217, 220, 223–24, 230, 241, 243–45, 295, 342
  'Gadget' 306
  Germany, development 76, 301–304
  heavy water source 142–43, 304
  Hiroshima and Nagasaki bombings 179, 294, 298, 307–309, 310, 312, 314, 320, 333, 342–43
  Japan, development 305
  'Little Boy' 307–309
  Manhattan Project 183–84, 222, 238–42, 247–48, 293–96, 298–301, 312
  MAUD Committee 142–44, 181, 184–85, 187–88, 189, 216, 217, 221, 223–24, 230–31, 233, 238, 239, 240, 243, 244–45, 292, 295
  moral issues 310, 312
  Oliphant contribution 6–7, 115, 134–35, 216–17, 235, 245–48, 309–10, 316, 321–24, 341–42

Index

origins, accounts of 241–48
plutonium 187, 229
Second World War, necessity arguments 310–14
scientist qualms 132–34
splitting the atom 68–81, 94
'Statement Relating to the Atomic Bomb' 246
test, first 305–306
United States' Official History of the Manhattan Project 242
uranium supply 183–84, 302
wartime race to produce 237–41
Atomic Energy Authority (Britain) 241
'Memorandum on the Properties of a Radioactive "Super-bomb"' 241
'On the Construction of a "Super-bomb"; based on a Nuclear Chain Reaction in Uranium' 241
Atomic Heritage Foundation (US) 342
Augusta (USS) 225
Australia
 contemporary success 348–49
 cultural cringe 247–48
 Federation 13, 14
Australian Academy of Science 332
Australian National University 30, 97, 317, 324–26, 331–32
 John Curtin School of Medical Research 326
 Research School of Physical Sciences 332
aviation technology 50–51

B

Bacillus pyocyaneus 88
Bacon, Francis 33
Bainbridge, Kenneth 184
Baldwin, Stanley 46–47, 61, 68, 104
 disarmament theory 46–47

Barcroft, Joseph 31
Barnes, Dr John 123
Barry, Hugh 150
Bartell, Professor Lawrence 247–48
Battle of Britain *see* Second World War
Baxter, James 264
 Scientists Against Time 264
Bazeley, Percival 284
Beaton, Cecil 32
Beaverbrook, Lord 275
Behring, Emil Adolf von 14
Bell Telephone Company 178–99, 215
Bell Laboratories 253
Bickel, Lennard 65, 199, 203, 338–39
Blackburn, Elizabeth 349
Blackett, Lord Patrick 232
Blunt, Anthony 38
Bohr, Niels 70, 78, 93–94, 117, 120, 126–28, 131, 143, 301, 305
Boot, Harry 100, 108, 109–15, 121, 125, 271
Bowen, Edward 'Taffy' 98–99, 170–72, 175–78, 214, 252, 261, 264
Bragg, Lawrence 15, 43–44, 110, 349
Bragg, William 15, 349
Briggs, Dr Lyman 180–81, 184, 187, 217–20, 230, 294, 303–304, 322
Britain 28
 air defence 52–53
 air raid planning 105–106
 Australian migration to 10, 15
 fear of war 46–49, 71
 Great Strike 1926 37
 interwar years 46–52
 radar development *see* long-wave radar; microwave radar

Second World War *see* Second World War
British Central Scientific Office, Washington 182
British Chemical Society 127, 129
British Empire 10, 12–13
British Journal of Experimental Pathology 87, 89
British Technical and Scientific Mission (Tizard Mission) 172–78, 180, 182–83, 184–85, 189, 212, 214–15, 252, 254
Brittain, Vera 48
Testament of Youth 48
Bronowski, Jacob 25
Browne, Maurice 68
Wings over Europe 68–69, 71, 78
Buderi, Robert 259
Bundy, McGeorge 216, 243, 245
Burgess, Guy 38
Burnet, Sir Macfarlane 337
Bush, Dr Vannevar 174–75, 181, 182, 184, 221, 223–24, 228, 230, 235–36, 238–39, 243, 254, 294, 295–96, 306, 316
Butcher, Lieutenant Commander Harry 266
Bynum, Bill 277

C

Cairns, Hugh 30, 279
Cambridge Philosophical Society 127
Proceedings 127
Campbell-Renton, Margaret 88
Carlsberg Laboratory, Copenhagen 117
Carnegie Institution for Science (US) 174, 223
Cephalosporin C 320
Chadwick, James 36, 43, 73–74, 173–74, 188, 299

Chain, Ernst Boris 84–90, 117–19, 121, 122–24, 150–51, 155, 160, 163, 164, 196–97, 269, 277, 288–89, 319, 327–28
Chamberlain, Neville 102–103, 105, 138, 149
Chariots of Fire 32
Chifley, Ben 326
Churchill, Winston 47, 48–49, 58, 60–61, 71–72, 137–39, 149–50, 153–54, 157–59, 164–65, 174, 182, 188–89, 211, 212, 225, 249, 265, 275, 279–80, 281–82, 294–95, 297, 343
CIBA 282
Clark, Manning 348
History of Australia 348
Clark, Ronald 241
Clusius, Klaus 129
Cockburn, Stewart 321, 324
Cockcroft, Sir John 36, 69, 70, 71, 73, 96–99, 142, 143, 171, 175–78, 184, 218, 322, 326
Coetzee, JM 349
Cold War 327, 335, 343
Compton, Arthur 233, 237, 238, 239, 246, 293
Compton, Karl 253
Conant, James 220–22, 228, 230, 233, 237–39, 243, 293, 306
History of the Development of an Atomic Bomb 222
Connaught Laboratories, Toronto 226
Coolidge, William 222–23
Coombs, Dr HC 'Nugget' 286, 324, 327
Cornforth, John 349
Cox, Johnny 194, 203
Craven, Peter 348
Crick, Francis 110
Curtin, John 283–84, 286

Index

D
Daily Express 149
Daily Telegraph 321
Dale, Sir Henry 195, 2898
Darwin, Charles 33, 339
Darwin, Sir Charles 182, 185, 217, 322
Dawson, Dr Martin 192, 227
Dean, Professor Henry 32, 33, 34
death rays 49–50, 53–54
Defries, Dr Robert 226–27
Desirade 297
Dimbleby, Jonathan 265
Discovery 79
Doenitz, Admiral Karl 250, 251, 252, 255, 256, 258, 265, 271
Doherty, Professor Peter 277–78, 349
Dowding, Air Marshal Sir Hugh 57, 165
Dreyer, Professor Georges 39, 88
Dreyfus, Catherine 202
Drury, Sir Alan 33–34
Dubos, Dr Rene 286–87
DuBridge, Lee 179, 215, 253, 315
Duchess of Richmond 172, 173, 174
Dunn, Sir William 82
Dunstan, Don 334

E
Ehrlich, Paul 38
Einstein, Albert 24, 31, 54, 70, 76–77, 180, 241, 244
Eisenhower, General Dwight D 266
electromagnetic waves 53
Ellis, Charles 43
Ellyard, David 324
Enola Gay 307–309, 314

F
Faulkner, William 343
Fermi, Enrico 74, 76, 79, 222, 230
Fine, Norman 264–65
First World War 19–20, 27, 47, 49, 62, 71
deaths from infection 64
Firth, Professor Raymond 324, 328
Fleming, Alexander 38, 62–63, 64, 66–67, 86–88, 122, 124, 162–64, 271–73, 287, 289, 319, 320
Fleming, Ian 201
Fletcher, Dr Charles 192
Florey, Bertha 19
Florey, Charles 35, 204, 275, 330
Florey, Ethel Hayter (nee Reed) 20, 30, 35, 86, 270–71, 273, 330
Florey, Hilda 17, 20, 285, 286
Florey, Howard Walter
  antibacterial agents research 38–39
  antibiotics research 320
  Australian National University (ANU) 324–26
  Canada 226
  Chain and 84–85, 196–97, 319
  children *see* Florey, Charles; Florey, Paquita
  contribution, assessment of 6, 235, 272–73, 277–78, 316, 337–39, 345–47, 349
  doctorate 31–32, 36
  family and childhood 16–17
  Fellowship 36
  Fleming and 162–64, 274–76, 319
  friendship with Oliphant 9–10, 11, 23, 317–18, 330–31
  fundraising role 90, 118–19, 120–21, 155–56, 192, 195–96, 277
  knighthood 284
  The Lancet 156, 160–63, 192, 226, 271, 274
  life peerage 329
  London Hospital Research Fellowship 35
  marriage 35

medical discoveries and 26–27
Nobel Prize 33, 84, 289–90, 318–20
North Africa 278 80
Order of Merit 329
penicillin research 88–89, 116–24, 144–48, 150–52, 154–56, 160, 190–97, 226–27, 236, 337–38
personality 30–31, 33–34, 85, 86, 275
Professorship of Pathology 36, 82–84
Provost of Queen's College, Oxford 329
Rhodes Scholarship/Oxford 9–10, 20–21, 27–31, 39
Rockefeller Foundation travel grant 34, 119
Royal Society President 29, 328–29
'salesman' role 3, 196–97, 202–206, 207–210, 226–27, 236, 269–70
Sir William Dunn School of Pathology 82–83 *see also* penicillin research *above*
Soviet Union visit 281
United States 34–35 *see also* wartime missions to US *below*
University of Adelaide 19, 285–86
University of Cambridge 31–33, 36
University of Oxford 9–10, 20–21, 27–31, 39 *see also* Sir William Dunn School of Pathology *above*
University of Sheffield 36, 38
war work in Pacific 283–84
wartime missions to US 2–3, 6–7, 198–210, 225–27
Westminster Abbey memorial stone 339
Florey, Joseph 16

Florey, Paquita 35, 88, 204
Florey, Valetta 285
Flugge, Siegfried 79
Foch, Ferdinand 52
Foote, Shelby 312
Fowler, Professor Ralph 182, 184, 185, 217
Franco, General Francisco 51
Frisch, Otto 77, 94–95, 120, 124–25, 127–33, 136, 143, 144, 186, 271, 297, 301, 305–306, 310, 321–22, 327
Fuchs, Klaus 327
Fulton, John 30, 204, 277, 288
Fulton, Lucia 204
Fussell, Paul 312

G

Gaddis, John Lewis 343
Gardner, AD 84, 160
General Electric & Co 116, 222
'germ theory of disease' 65
Germany 28, 37
  atomic bomb development 301–304
  Luftwaffe 51, 99, 103–104, 106, 118, 140–41, 144, 156, 159–60, 164–66, 173, 250, 253, 259
  radar development 57–58
  Second World War *see* Second World War
  Wehrmacht 140, 141, 157, 159, 211–12, 234, 281
Goering, Hermann 141, 159, 259–60
Gordon, Professor George 120
Gort, General Lord 140
Gowing, Margaret 187
Grant, Professor Kerr 22, 23
The Great Artiste 179
Great Depression 47, 51, 214
Gregg, Dr Alan 203

Index

Groves, Lieutenant General Leslie 242, 298, 306, 321, 323
Guderian, General Heinz 141

H
Hahn, Otto 76, 77, 80, 93–94, 302
Haldane, Jack 31, 33
Halifax, Lord 138–39
Hancock, Keith 30, 105, 167–68, 324, 325, 328
  Australia 30
Hare, Dr Ronald 226
Hargadon, Andrew 272–73, 278
Harris, Sir Henry 116–17, 210, 246, 278
Harrison, Peter 285
Hartmann, Kurt 103
Hastings, Baird 208
Hawkins, Percy 194
Haworth, Norman 144
Heatley, Norman 2, 85–86, 116–18, 122–23, 147–48, 150–52, 155, 160, 163, 190–92, 195, 196, 203, 206–207, 225, 227–28, 252, 269–70, 277, 290
Heisenberg, Werner 301, 302–304
Herrick, HT 228
Herschel, John 339
Herschel, William 339
Hertz, Heinrich 55, 111
Heydrich, Reinhard 200
Hill, Professor AV 181–82, 184, 185
Hiltzik, Michael 238
Hiroshima and Nagasaki bombings 179, 294, 298, 307–309, 310, 312, 314, 320, 333, 342–43
Hitler, Adolf 1, 2, 104, 105, 106, 133, 141, 158, 159, 166, 211, 258, 283, 301, 304
Hooke, Robert 46
Hulsmeyer, Christian 55
Humphries, Barry 348

Hunt, Mary 267–68, 284
hydrogen bomb 42

I
Imperial Chemical Industries 279, 292
Imperial College (Britain) 127, 142
Institute of Theoretical Physics, Copenhagen 117, 120
Ironside, General Sir Edmund 140, 154
Isherwood, Christopher 32–33
Ismay, General Hastings 'Pug' 150, 153

J
Japan *see* Second World War
Jennings, Dr Margaret 86, 152, 331
Jewett, Frank 223
Joliot-Curie, Frederic 79, 183
Jones, RV 304

K
Kellaway, Sir Charles 65–66
Kennedy, Joe 137–38
Kent, James 36, 146–47
Khariton, Yuri 37
King George V 48
King George VI 149, 284
Kistiakowsky, George 220
Koch, Robert 26, 29
Korda, Alexander 49
  Things to Come 49
Kretchmar, HH 284
Ladenburg, Rudolf 303–304

L
Lambert, Harry 273
The Lancet 156, 160, 192, 271, 274
  'Further Observations on Penicillin' 226
  'Penicillin as a chemotherapeutic agent' 160–63, 192

Large Hadron Collider 323
Lawrence, Ernest 91–92, 175,
 177, 178, 179, 187, 229–33, 237,
 239, 240, 246, 253, 291, 293,
 298–300, 305, 321, 332
Lawrence, Gunda 92
Lawrence, John 92
LeMay, General Curtis 262–63
Lederle 235
life expectancy 63–64
 antibiotics and 337
Liljestrand, Dr Göran 289, 319
Lindemann, Frederick (Lord
 Cherwell) 188, 221
Lisbon 198–201
Lister, Joseph 26, 29, 272
Loomis, Alfred 175, 176–79, 182,
 264
long-wave radar
 Bawdsey Manor 57, 60–61, 99
 Chain Home network 58–59, 97,
  99–100, 103, 116, 165–66, 255
 'The Daventry Experiment' 56, 61
 development 55–57, 97–98
 Germany 57–58
 long-wave radio waves 59
 origins 54–55
 research 54–55, 60

M
Macdonald, HM 110–11
 Electrical Waves 110–11
Macfarlane, Gwyn 118, 330
McGill University, Montreal 14
Makinsky, Alexander 199
Malouf, David 348
Marconi, Guglielmo 14, 50, 55
Marshall, Barry 349
Massachusetts Institute of
 Technology's Radiation
 Laboratory (Rad Lab) 179–80,
 215, 253, 261, 315

Massey, Harrie 36, 93
Matthews, Harry Grindell 50
MAUD Committee *see* atomic bomb
Maxwell, James Clerk 54–55
 *Treatise on Electricity and
  Magnetism* 54–55
May, Alan Nunn 327
May, Dr Orville 205–206
Medical Research Council (Britain)
 90, 118–19, 156, 192
Meitner, Lise 77, 94, 120, 126, 143
Mellanby, Sir Edward 90, 118, 119,
 122, 156, 210, 275, 276, 287, 330
Merck 235, 269
microwave radar 4, 6, 54, 224, 250,
 252, 340–41
 'Blind Bombing Radar' 254
 cavity magnetron 4, 109–16, 121,
  125, 144, 168–69, 171, 177–80,
  212, 214–15, 254, 263–65, 271,
  285, 297, 315–16, 322, 340–41
 Germany 258–61
 High Frequency Direction Finding
  Equipment (HF/DF) (Huff-
  Duff) 255
 H2S ground-mapping radar 254,
  257
 Japan 260–61
 klystron 108–110, 112, 114, 168
 'The Magnetron Memorandum'
  112, 131
 mass-produced 253
 microwave technology 59–60,
  98–99
 Oliphant contribution 6–7,
  235, 250, 264–65, 309–10, 316,
  340–41
 research 97–101, 108–15, 124–25
 Second World War contribution
  260–65, 293, 315–16, 340
 'strapped' magnetron 215
 submarines, use against 255–58

Index

Miller, Harry M 'Dusty' 119–21, 195
Montgomery, Field Marshal Bernard 256
Moon, Philip 142
Moran, Lord 275
Morell, Theodore 283
Moyer, Dr Andrew 206, 225, 227–28, 269, 287
National Academy of Sciences (US) 35, 223, 239–40, 294, 295
National Bureau of Standards (US) 180
National Defense Research Committee (NDRC) (US) 175, 181, 184, 208, 220
  Microwave Committee 175, 176, 178

N
National Library of Australia 338
National Physical Laboratory (Britain) 53
  Radio Department 53
Nature 22, 94
Naval Research Laboratories (US) 253
Necessary Evil 314
Nernst, Walter 71
New York Times 202
Newton, Isaac 24, 33, 45–46
Nichols, Robert 68
  Wings over Europe 68–69, 71, 78
Nicolson, Harold 149, 158
Nishina, Yoshio 305
Nobel Institute of Medicine 319
Nobel prizes 14–15, 33, 41, 84, 229, 253, 288
  Australian recipients 348–49
  ceremony 1945 318–19, 320
  chemistry 14

medicine 14, 33, 84, 277, 278, 288–89, 318–19
physics 74, 179
Norness 251
Norsk Hydro power plant, Norway 142–43
Northern Regional Research Laboratory (NRRL) *see* US Department of Agriculture
nuclear energy 344
Nuffield, Lord 91
Nuffield Research Laboratory (Britain) 113, 167

O
Office of Scientific Research and Development (OSRD) (US) 175, 208–209, 215, 224, 235, 280
  Committee on Medical Research 35, 208, 234–35, 287
Oliphant, Beatrice 18
Oliphant, Geoffrey 41, 335
Oliphant, Harold ('Baron') 16
Oliphant, Keith 18
Oliphant, Mark (Marcus) Elwin
  Australia, radar assistance 296–97
  Australian National University (ANU) 317–18, 324–26, 328, 331–33
  Chifley meeting 326–27
  Cold War victim 327
  contribution, assessment of 6–7, 115, 134–35, 216–17, 235, 245–48, 250, 264–65, 309–10, 316, 321–24, 340–42, 345–47, 349
  cyclotrons 91–92, 96, 108, 167, 229, 298, 300, 323, 328
  doctorate 41
  Ernest Lawrence and 91–92, 229–33, 237–38, 297–301

419

Exhibition Scholarship 36
family and childhood 16–20
friendship with Florey 9–10, 11, 23, 317–18, 330–31
klystron *see* microwave radar
nuclear weapon views 333–34
physics, love of 25, 41–42, 94, 131
proposed Medal of Freedom with Gold Palm award 321–23
public speaking 333
public recognition 321
radar *see* microwave radar
role in atomic bomb development, assessment 243–47
South Australia's governor, as 334–35
synchrotron 323–24, 328
University of Adelaide 21–23
University of Birmingham 30, 38, 43–44, 60, 90–95, 107–108
University of Cambridge 10, 36–43
war work/military research 96–100, 107–109, 112–15, 142–44, 173–74, 291–92, 344–45
wartime missions to US 4–6, 210–24, 228–33
Oliphant, Michael 144, 335
Oliphant, Rosa (nee Wilbraham) 23, 90–91, 144, 296, 334, 335
Oliphant, Vivian 144
Oppenheimer, Robert 232–33, 240, 298, 306, 310, 316, 321, 334
Orr-Ewing, J 160
Osler, William 26

P

Park, Air Vice Marshal Sir Keith 164
Pasteur Institute 282
Pasteur, Louis 19, 26, 66, 272
Peierls, Genia 95

Peierls, Rudolf 93, 95, 124–27, 129–33, 136, 186–87, 271, 293, 297, 301, 306, 322, 327
penicillin 3, 30, 32, 34, 62–67, 65
 Australian manufacture 285
 availability 280–81
 cost 207–209, 280
 credit for discovery 269–70, 271–78, 286–90, 319
 'deep-tank fermentation' method of production 225–26, 268
 discovery 66–67
 drug companies and 235–36
 experiments 152
 German manufacture 282
 growth medium 122
 human trials 154, 192–95, 270–71
 instability 122, 235
 The Lancet articles 160–63, 192, 198, 226
 lives saved by 337
 manufacturing costs 207–209
 media attention 273–75
 moulds 65–66, 160, 225–26, 235, 277, 279
 Oxford memorial 336
 patents 269
 'Penicillin Girls' 192
 Penicillium chrysogenum 268
 Penicillium notatum 2–3, 65–67, 87–89, 116–17, 122–23, 155, 190–92, 205–207, 210, 225–26, 266–68, 271–72, 282
 production/yield challenges 154–55, 122–23, 162, 190–91, 193–94, 206–207, 225, 266
 'protective' mice experiment 145–48, 151, 156, 161
 research 87–89, 116–22, 144–48, 150–51
 reverse extraction 122–23
 testing 123–24

Index

Petain, Marshal Phillipe 157
Pfizer 207, 235
Philby, Kim 38
Phillips, AA 347–48
Picasso, Pablo 51
Pincher, Chapman 243–44
Popov, Dusko 201
Prince of Wales (HMS) 225
Pugwash Conferences on Science and World Affairs 334
Pulvertaft, Lieutenant Colonel Robert 279, 282

Q
Queen Elizabeth 149
Queen Victoria 13

R
Rabi, Isidor 25, 179, 243, 252, 315–16
radar *see* long-wave radar; microwave radar
radiation therapy 92
radio waves 53–54, 111 *see also* long-wave radar; microwave radar
Rad Lab *see* Massachusetts Institute of Technology's Radiation Laboratory (Rad Lab)
Ramsey, Norman 179
Randall, John 100, 108, 109–15, 121, 125, 271
Raytheon 253, 341
resonant cavity magnetron *see* microwave radar
Reynaud, Premier Paul 149
Reynolds Illustrated News 69
Rhodes, Cecil 20
Rhodes, Richard 246, 340, 344
  The Making of the Atomic Bomb 246
Richards, Dr Alfred Newton 35, 208–210, 234, 236, 266, 277
Robertson, Brailsford 21

Rockefeller Foundation 34–35, 195–96, 199, 202–204, 208, 287
Rockefeller Sr, John D 120
Rommel, Field Marshal Erwin 256
Roosevelt, President Franklin D 138, 174–75, 180, 217, 223, 224, 225, 238–39, 241–43, 281, 294–95, 297, 299
Roskill, Captain SW 257–58
Rowe, AP 'Jimmy' 52, 315
Royal Air Force (RAF) (Britain) 48, 52, 106, 134, 159, 164, 173, 250
Royal Caroline Institute (Sweden) 319
Royal Navy (Britain) 152–53, 174, 255
Royal Society (Britain) 29, 33, 195, 271, 328–29
Runciman, Steven 32
Rutherford, Sir Ernest 5, 14, 22–23, 24, 29, 31, 36, 40–43, 69–70, 71, 90, 96, 182, 292, 332

S
Sachs, Alexander 180
Sackville-West, Vita 149
St Mary's Hospital Medical School, London 62, 66, 273, 274, 275
Salazar, Antonio de Oliviera 199
Sanders, AG 160
Sassoon, Siegfried 51
Sayers, James 215
Schellenberg, Walter 200–201
Schmidt, Professor Brian 349
science fiction 49–50
scientific espionage 37–38
Scott, Robert 107
Seaborg, Glenn 187
Second World War 102–106, 173, 189
  air raids in Japan 262–63
  American entry 240–41, 251

Atlantic Charter 225
atomic bomb *see* atomic bomb
Battle of the Atlantic 187–88, 212, 249–52, 256, 257–58, 264, 265
Battle of Britain 2, 164–67, 173, 250
Battle of the Philippine Sea 261–62
Birmingham Blitz 107, 167
blitzkrieg 139–40, 211
British Expeditionary Force (BEF) 139–40, 144, 149, 152–53, 156–57
declaration of 61, 100, 102–103
Dunkirk 141, 144, 149, 152–54, 156
Enigma Code, breaking of 257
French Armistice 157
German surrender in Stalingrad 256
'The Greenland Gap'/'The Black Pit' 251, 252, 256, 257
Japanese surrender 309, 314–15
London Blitz 159, 163, 259
Luftwaffe 51, 99, 103–104, 106, 118, 140–41, 144, 156, 159–60, 164–66, 173, 250, 253, 259
Maginot Line 139
military-scientific quests 45–46
North African campaign 212, 256–57, 278–79
Norwegian campaign 137–38
nuclear weapons 4–6 *see also* atomic bomb
Operation Downfall 313
Operation Drumbeat 251
Operation Sea Lion 158–59
Operation Typhoon 234
Pacific theatre 261–63, 283–84, 293
Pearl Harbor 240, 294, 305
penicillin discovery *see* penicillin
Phoney War 106, 136, 137
Poland, invasion of 61, 100, 103–104
radar, impact 260–65, 293, 315–16, 340 *see also* long-wave radar; microwave radar
scientific information, exchange of 1–6, 170–78, 180–85, 214–15
Soviet Union, invasion of 3, 189, 211–12
Tehran meeting of Allied leaders 281
Wehrmacht 140, 141, 157, 159, 211–12, 234, 281
Segre, Emilio 187
Sengier, Edgar 183
Shellenberger, Michael 344
Sherrington, Sir Charles 29, 31, 33, 35, 270, 273
Shirer, William 105
Simon, Franz 186–87
Slessor, Air Marshal Sir Jack 257
Smith, John 207
Snow, CP 79–80, 264
Soviet Union
 declaration of war on Japan 311
 Florey visit to 281
 invasion by Germany 3, 189, 211–12
 scientific espionage by 37–38
Spanish Civil War 51
Spanish flu 64
Speer, Albert 283, 304
Spencer, Percy 341
Squibb 235
Stalin, Joseph 281
Standard Oil 202
Stanford University 108
Stapledon, Olaf 49
 Last and First Men 49
Stauffenberg, Claus von 283
Stimson, Henry 178

Index

Strassman, Fritz 76, 77, 80, 94
Subtilis-mesentericus 88
Szilard, Leo 75–76, 77, 79, 180, 219, 246–47

T

Taylor, AJP 2–3, 257
Telecommunications Research Establishment (Britain) 254
Teller, Edward 180
Tennyson, Lord 11
Tesla, Nikola 50, 55
Theorell, Professor AHT 319, 320
Thom, Dr Charles 205, 228, 235
Thompson, Inspector WH 139, 225
Thomson, George 127, 142, 183, 185–86, 189, 216, 239
Thomson, JJ 14, 31, 127, 232
Tibbets, Colonel Paul 307, 308
Time magazine 309
The Times 273–74
Tizard Committee (Committee for the Scientific Survey of Air Defence) 53, 61, 135
Tizard Mission (British Technical and Scientific Mission) 172–78, 180, 182–83, 184–85, 189, 212, 214–15, 252, 254
Tizard, Sir Henry 53, 97–99, 125, 135–36, 142, 174, 175, 178, 183
Truman, President Harry 321
Tube Alloys 188, 246, 292–93, 295, 298, 301
Tuck, James 229
Twain, Mark 12
twentieth century scientific and technological advances 14–16
  acceleration through war 45–46
  atomic bomb *see* atomic bomb
  aviation technology 52–53
  Big Science 345–46

Florey contribution *see* Florey, Howard Walter
medical advances 25–26, 62–64, 92 *see also* penicillin
Oliphant contribution *see* Oliphant, Mark (Marcus) Elwin
physics 24–25
radar development *see* long-wave radar; microwave radar
Tyndall, John 272

U

Udet, Ernst 57–58
UK Atomic Energy Commission 134
Union Miniere du Haut Katanga 183
United Nations Atomic Energy Commission 327
United Nations Charter 225
University of Adelaide 9, 15, 17, 19, 30, 39
University of Birmingham 4, 30, 43–44, 60, 90–91, 107–108, 144, 327
University of Bristol 60, 107
University of California 91–92
University of Cambridge 31–33, 36–37
  Cavendish Laboratory 22–23, 36–37, 39–42, 90, 96, 110, 142, 232, 299, 332
University of Chicago 233
University of Copenhagen 93
University of Oxford 9, 27–29, 120
  Clarendon Laboratory 60, 186
  Sir William Dunn School of Pathology 39, 82–85, 89–90, 117–18, 120, 146, 155, 163, 190, 195, 210, 270, 275, 277, 278, 325
University of Sheffield 36, 38
University of Sydney 261
  Radiophysics Laboratory 261

US Department of Agriculture Northern Regional Research Laboratory (NRRL) 205–207, 228, 236, 267, 287

V
van der Waerden, Bartel Leendert 303
Varian, Russell 108
Varian, Sigurd 108

W
Wake, Nancy 323
Wallace, Vice President Henry 238
Walter and Eliza Hall Institute, Melbourne 65
Walton, Ernest 36, 69, 70, 71, 73, 96
Washington, President George 25
Watson, James 110
Watson-Watt, Robert Alexander 53–54, 56–59, 61, 98, 112, 165, 167, 179, 255, 341
Waugh, Evelyn 28
Brideshead Revisited 28

Weaver, Warren 196, 202, 227
Weir, Sir Cecil 276
Wellcome Research Laboratories (Britain) 155–56
Wells, HG 49, 104
Western Electric 253
Wheeler, John 126, 127–28
White, Patrick 347
'The Prodigal Son' 347
Wigner, Eugene 180
Wilkins, Arnold F 'Skip' 53–54, 56
Williams, Dr Trevor 272
Wilson, AN 6
Wimperis, HE (Harry) 52–53
Wordsworth, William 33
Wright, Charles 107
Wright, Dr Roy Douglas ('Pansy') 88–89
Wright, Sir Almroth 62, 274

Y
Yasuda, Lieutenant General Takeo 305